# THE SPACE TRANSPORTATION MARKET:
## EVOLUTION OR REVOLUTION?

# SPACE STUDIES

## VOLUME 5

Editor
Prof. MICHAEL RYCROFT

**International Space University**
**Excellence in space education for a changing world**

The International Space University (ISU) is dedicated to the development of outer space for peaceful purposes through international and interdisciplinary education and research. ISU works in association with a number of Affiliates (universities, research institutes, consortia ...) around the world and in partnership with space agencies and industry.

For young professionals and postgraduate students, ISU offers an annual ten-week Summer Session in different countries and a one-year master of Space Studies (MSS) program based at its Central Campus in Strasbourg, France. ISU also offers short courses and workshops to professionals working in space-related industry, government and academic organizations.

Independent of specific national and commercial interests, ISU is an ideal forum for discussion of issues relating to space and its applications. The network of alumni, faculty, guest lecturers, Affiliate representatives and professional contacts which characterizes the ISU Community makes it possible to bring together leading international specialists in an academic environment conductive to the exchange of views and to the creation of innovative ideas. ISU aims to promote productive dialogue between space-users and providers. In addition to the Annual Symposium, ISU supports smaller forum activities, such as workshops and roundtables, for constructive discussions which may help to chart the way forward to the rational international utilization of space.

# THE SPACE TRANSPORTATION MARKET: EVOLUTION OR REVOLUTION?

Edited by

M. Rycroft

*International Space University, Strasbourg, France*

KLUWER ACADEMIC PUBLISHERS

DORDRECHT / BOSTON / LONDON

A C.I.P. Catalogue record for this book is available from the Library of Congress.

ISBN 0-7923-6752-9

Published by Kluwer Academic Publishers,
P.O. Box 17, 3300 AA Dordrecht, The Netherlands.

Sold and distributed in North, Central and South America
by Kluwer Academic Publishers,
101 Philip Drive, Norwell, MA 02061, U.S.A.

In all other countries, sold and distributed
by Kluwer Academic Publishers,
P.O. Box 322, 3300 AH Dordrecht, The Netherlands.

*Printed on acid-free paper*

Printed in the Netherlands

ISU gratefully acknowledges the cooperation of the American
Astronomical Society (AAS) and the financial support provided by


**The Boeing Company**

**Centre national d'études spatiales (CNES)**

**European Space Agency**

**Lockheed Martin Space Operations**

**New Skies Satellites N.V.**

**United Space Alliance**

# Table of Contents

# Acknowledgements

ISU acknowledges with thanks the advice and support given by the following people as members of the Program Committee:

**R. Akiba**, Space Activities Commission, Science and Technology Agency, Japan

**A. M. Browne**, Chief Financial Officer, New Skies Satellites N.V., The Netherlands

**F. Engström**, Director of Launchers, ESA

**A. van Gaver**, Senior Technical Advisor, Directorate of Launchers, ESA

**J. F. Honeycutt**, President, Lockheed Martin Space Operations, USA

**A. Kerrest**, Faculty of Law, University of Western Brittany, France

**R. Khadem**, Chief Financial Officer, INMARSAT, United Kingdom

**I. Pryke**, Head of ESA Washington Office

**G. Laslandes**, Deputy Director, Space Transportation and in Orbit Infrastructures, CNES-HQ, France

**G. Schluter**, VP and General Manager, Expendable Space Systems, The Boeing Company, USA

**R. Stephens**, VP and General Manager, Reusable Space Systems, The Boeing Company, USA


Symposium Programme Chair and Convenor: **Y. Fujimori**, ISU

Symposium Co-ordinator: **L. Chestnutt**, ISU

Proceedings Editor: **M. Rycroft**, ISU

Editorial Assistant: **N. Crosby**, ISU

# Foreword

**M. Rycroft**, Faculty Member, International Space University

e-mail: Rycroft@isu.isunet.edu

"The Space Transportation Market: Evolution or Revolution?" was the question which was the focus for the papers presented, and also the Panel Discussions, at the fifth annual Symposium organised by the International Space University. Held in Strasbourg, France, for three lively days at the end of May 2000, the Symposium brought together representatives of the developers, providers and operators of space transportation systems, of regulatory bodies, and of users of the space transportation infrastructure in many fields, as well as experts in policy and market analysis.

From the papers published here, it is clear that today's answer to the question tends more towards evolution than to revolution. The space launch industry is still not a fully mature one, and is still reliant on at least partial funding by governments. Better cooperation is essential between governments, launch providers, satellite builders and satellite operators in order to reduce the problems which the space transportation market faces today.

Among the 150 attendees were all members of the fifth Master of Space Studies class, young professionals and postgraduate students who are developing an aspect of the Symposium's theme in their Team Project. Their final report will be completed at the end of July 2000, and published separately.

# Keynote Address

# Socio-Economic Considerations of Future Space Transportation Systems

**R. Akiba,** Space Activities Commission of Japan, 2-2-1, Kasumigaseki, Chiyoda-ku, Tokyo 100-8966, Japan

e-mail: akibarsp@ma4.justnet.ne.jp

**S. Nomura,** National Space Development Agency of Japan, 2-4-1, Hamamatsu-cho, Minato-ku, Tokyo 105-8060, Japan

email: Nomura.Shigeaki@nasda.go.jp

**Abstract**

The present average reliability of launchers, about only 95%, is clearly insufficient to satisfy the needs of space users. Further essential improvements of reliability will be achieved by the development of fully reusable systems, though this does not exclude the use of expendable systems in parallel. Future human society will feature large populations, limited resources, and pollution originating from human activities. Consequently safety and reusability are mandatory issues in designing future systems. Economically, space transportation costs and demand are interdependent, an inverse relation being the marginal case which keeps total sales revenue constant. Technological efforts in a competitive market can reduce the costs of space transportation to some extent, and will show an evolutionary increase in the market. However, a revolutionary change in demand will come only after a mass transportation system with high reliability has been developed. So, steady progress toward those systems is a key to opening the new era of space utilization.

## 1. Introduction

Various research and development efforts for future space transportation systems have been progressing internationally. However, differences in their background philosophies result in many different configurations. The present paper addresses the importance of socio-economic considerations to settle a goal for developments in the long-term, for instance ten to fifteen years ahead.

First of all, looking at the present status, the environment for space transportation systems is quite harsh. As Japan suffered two successive failures recently, the present reliability of launch vehicles is as low as 95 %. Some customers complaining at the low reliability of launchers find a way out of their difficulties by insurance. But, the lost time and efforts cannot be recovered fully by insurance. Time is not really money! Why is the present reliability so low? The answer to the question seems to be found in the development history of space launch vehicles. Their debut was a missile known as V-2 in which low

1

*M. Rycroft (ed.), The Space Transportation Market: Evolution or Revolution?, 1–8.*
© 2000 *Kluwer Academic Publishers. Printed in the Netherlands.*

reliability was acceptable for its purpose. The following age of further developments was dominated by the cold war in which, again, reliability was not given first priority — except for manned flights such as the Apollo project. Notwithstanding the effort enhancing reliability in that age, it was still too low for civilian use.

Historically, a big challenge to improve this situation was the development of the Space Shuttle. It might have been a chance to raise the reliability of space transportation systems (STS) since its reusability could have been effective to reduce initial failure modes. But, unfortunately, this was not well realized as it did not adopt a fully reusable design. Everyone remembers the Challenger tragedy.

It is quite natural that NASA's new budget emphasizes an enhancement of the safety of STS. The solution to improve the present situation is clear — that is the development of a fully reusable launch vehicle. Today's question is why there is no project for a genuine reusable launcher in the world, and another is how we should proceed towards that.

## 2.    Future Human Society

In order to settle the target of development of space transportation systems in future, we need to pay attention to the environment of future society. First, we should recognize that we are coming to a turning point in human history. The rapid development of modern society has brought about a population explosion.

According to recent statistics [Reference 1] the world population amounts to 5849 million, that is almost 6 billion people, while the world surface area is approximately 510 million $km^2$. So the average population density has already reached 11 people per square kilometer. If we restrict ourselves to take only the land surface into account, this rises to 38 / $km^2$. Since the desert occupies approximately 15 % of total land area, 50 / $km^2$ seems to be a reasonable figure for the population density in habitable areas (see Fig. 1 [Reference 2]).

Accordingly, it is a natural consequence that the construction of new facilities which may disturb the surrounding environment often generates objections from environmental protection groups. So, there are only limited sites for launching space vehicles in the world, even at the present time. Social sensitivity to environmental problems will increase more and more in the future. Environmental problems include not only local but also global ones:

besides noise, pollutant emission, space debris, radiation hazards and other unknown effects of electromagnetic fields are typical examples (Fig. 2).

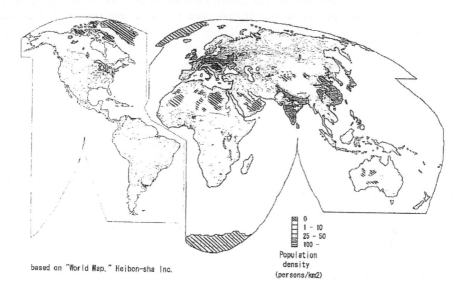

based on "World Map," Heibon-sha Inc.

0
1 – 10
25 – 50
100 –
Population density (persons/km2)

**Figure 1.** Population density of the world

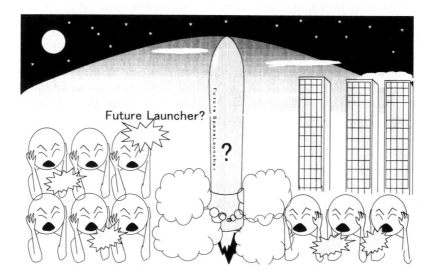

**Figure 2.** Is this the space transportation in the future?

There are not many launch sites among the space-faring nations. Taking into account the future increase in launch frequency and about the same number of recovery operations, it is too optimistic to assume that conventional launch systems can survive in our future society. Only airport-like operations will be acceptable to the public in the vicinity of launch sites. In other words, the noise and hazard levels should be kept as low as those of present airports. Of course, deserts are attractive as less populated areas.

Here, the authors emphasize the usefulness of the desert in the future. Limited resources and environmental problems will be social concerns spreading into our daily lives more and more in future. Among them, the energy problem is a vital issue for future human society.

As a renewable energy source, the Solar Power Satellite (SPS) system may offer an essential solution to that problem. One of the drawbacks of SPS is its thin energy density even at the power receiving area on the Earth, so it requires a large area for the receiving power at the Earth. The most realistic candidate sites will be found in the deserts. The total area of deserts amounts to approximately 22 million km$^2$. If we assume 1 kW per capita as the demand in the future, 6 billion people need 6 TW in total. Based on the assumption of 100 W/m$^2$ energy density at the receiving site, it requires a rectenna area of 60 thousand km$^2$, that corresponds to about 0.3 % of the total desert area on the Earth.   It is an area not a negligible size, because we must provide more than ten times the area for safety. Since there are many other future demands which will need large space on the Earth's surface, we must preserve the deserts carefully as a precious resource of humankind.

The Sea Launch system is attracting commercial attention, but its poor accessibility from industrialized areas seems to restrict its expanding use in future. In summary, we can conclude that only the use of airport-like environments can supply the launch operations for future space transportation systems.

## 3.    Economic Considerations

Cost reduction is the primary purpose of present development efforts in the international community of commercial space transportation. Suppliers are expecting an expansion of the market, but excessive competition for sales is the currently real situation of the market. However, cheaper costs will bring more demand [Reference 3], and vice versa [Reference 4]. As the result of an interactive process, the supply costs must balance with the size of demand eventually.   For the long range prediction of the costs, we should not overlook the advancement of technology. As a result of both the above processes, the modern personal computer costs only 1/10,000 of the cost of an Apollo-age computer. Generally, the long-range prediction of cost and demand is not very reliable (see Fig. 3).

Thus, we can simply estimate an eventually balanced cost with demand if we assume that the total sales of the expanding market never decrease. Based on a simple inversely-proportional relation, one-tenth of the cost brings ten times the demand. In other words, the present launch frequency of 150 per year will increase to 1500 per year, eventually, when the present launch cost of $ 10,000 per kg falls to $ 1,000 per kg. No one can accurately answer the question: when will those equilibrium values be reached? (see Fig. 4). Twenty years might be necessary before reaching an equilibrium with a transportation cost of $ 1,000 per kg, because a rather optimistic estimation indicates a tripling of the market after ten years as shown in Fig. 3. However, the expansion process of the market is not always evolutionary through commercial competition, but can also be revolutionary by a novel demand such as the SPS. Schematically explaining, those two expansion processes are shown in Fig. 5. Since the revolutionary process needs a high risk high return investment, the governmental support is essential to proceed the research and development of the new space transportation system.

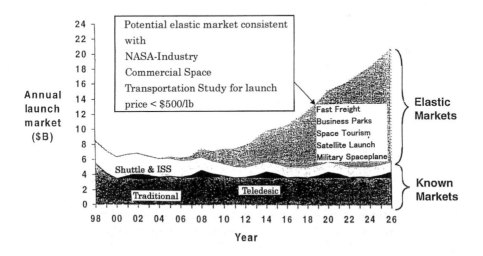

**Figure 3.** Estimated future launch markets

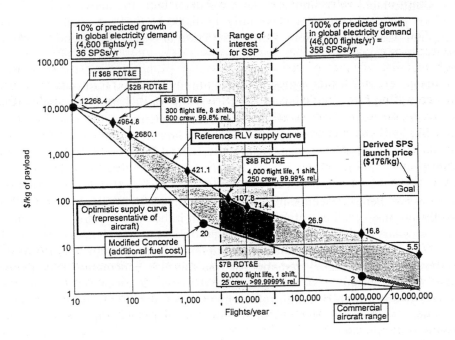

**Figure 4.** Supply costs for reusable launch systems

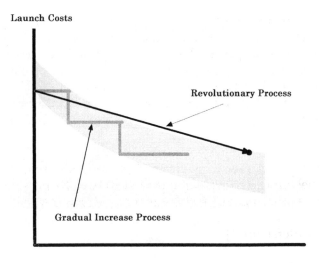

**Figure 5.** Schematic explanation of launch market expansion processes

## 4.    Propulsion System and Vehicle Configuration

The use of air-breathing propulsion is a logical consequence to satisfy an environmental condition that is acceptable to the public, since the inherently high exhaust velocity of rocket engines is a prohibitive source of noise in populated areas. The aerospace plane is an ideal configuration if it is realistic. But, according to a simple treatment [Reference 5], the use of an air-breathing engine seems to be effective for flight velocities up to Mach 10. The Air Turbo Ram (ATR) [Reference 6] is the most realistic air-breathing system in the near future. So, we propose Two Stage to Orbit (TSTO) configuration as the most appropriate configuration for future space transportation systems. Cost comparisons of TSTO with other candidates in future are not really meaningful because the future costs are quite ambiguous, as discussed in the previous section.

We should also note that the development of the air-breathing propulsion system for the aerospace plane is useful for future long-range aircraft capable of rapidly connecting two hemispheres without refueling. The transport capability of the first stage will be useful to improve the mobility of the upper stage to an arbitrarily convenient location. Since the TSTO operation makes the turn-around time shorter for the first stage, it will also bring an effect of cost reduction. The use of an air-breathing system in the launch phase will also increase the safety at the launch site, because the amount of explosive gas mixture is quite low in air-breathing systems.

The choice of fuel must also take environmental effects into account. Accordingly, the exhaust gas should contain only species existing in the natural atmosphere. Thus hydrogen fuel is the most promising candidate for both stages. If the upper stage employs an expendable vehicle, a cheaper vehicle is preferable. The hybrid propulsion system is an attractive candidate, besides the conventional storable liquid propellant rocket. The hybrid propulsion system has merits — it is cheap, safe, pollution-free, inherently restartable, throttlable and reusable.

Innovation of design is necessary to improve combustion efficiency and to increase the fuel burning rate. Several new ideas have been tested so far to give solution to those problems. Some of these unconventional hybrid rockets use:

1.)    Swirl flow [Reference 7]

2.)    Jet impingement [Reference 8]

3.)  Staged combustion [Reference 9].

If we find some effective design, the hybrid rocket can be useful for both reusable and expendable vehicles.

## 5.  Conclusion

The most promising reusable system is a two-stage launcher which consists of a jet airplane using ATR to accelerate it to an extremely high Mach number as the first stage, and a second stage rocket-plane. Use of an air-breathing system in the launch phase gives solutions to adverse effects in the vicinity of the launch site such as high noise, safety, and frequent takeoffs for mass transportation to space. The air-breathing first stage is also effective as a high speed aircraft in the atmosphere. Increased freedom for launch locations with the choice of different runways will also contribute to an expansion of demand.

The idea of TSTO proposed here is not new; most recently the Saenger project was interrupted due to economic difficulties. No private enterprise will invest in such a long-term project. Thus, only governmental efforts can sustain it. On the other hand, international cooperation is mandatory. Whatever the cooperation scheme may be, the culture of international technocracy is indispensable for steady development in an open environment.

### References
1.   The Japanese Astronomical Observatory, *Chronological Scientific Table*, 1999
2.   *World Map*, Heibonsha Inc., Tokyo, Japan, 1965
3.   Penn, J.: *Synergy Between Solar Power and Low-cost High Rate Reusable Launch*, IAF-98-R.2.09, 1998
4.   Andrews, D.: *Space Transportation Systems and Missions for Preparation of Propulsion Requirements*, CR90118-Reports of Advanced Propulsion Workshop #3, IAF Congress, Melbourne, 1998
5.   Akiba, R.: A Simple Theory of Aerospaceplane (in Japanese), *Journal of the Japan Society for Aeronautical and Space Sciences*, Vol. 37, No. 423, pp. 202-204, 1989
6.   Tanatsugu, N.: *Development Study on Air Turbo-ramjet*, Developments in High-Speed-Vehicle Propulsion Systems, 1st ed. Progress in Astronautics and Aeronautics, Vol. 165,  AIAA, Reston, pp. 259-331, 1996
7.   Yuasa, S.: *Effects of Swirl on the Stability of Jet Diffusion Flame, Combustion and  Flame*, Vol. 66, pp. 181-192, 1986
8.   Nagata, H., Okada, K., Sanda, T., Akiba, R., Satori, S. and Kudo, I.: *New Fuel Configurations for Advanced Hybrid Rockets*, 49th International Astronautical Congress, IAF-98-S.3.09, 1998
9.   *Incinerator Type Hybrid Rocket*, Library of Virtual Space Research Laboratory, Center for Advanced Science and Technology, Hokkaido University <http://gaea.cast.hokudai.ac.jp/vsrl/index.htm>.

# Session 1

# What will Constitute the Future Market for Space Transportation Services? Users' Perspectives

Session Chair:

**P. Smith,** FAA Office of Commercial Space Transportation, USA

# Commercial Space Transportation:
# Recent Trends and Projections for 2000-2010

P. Smith, U.S. Federal Aviation Administration Office of Commercial Space Transportation, 800 Independence Avenue, SW, Washington, DC 20591, USA

e-mail: patti.smith@faa.gov

### Abstract

Over the past 10 years, the number of commercial space launches worldwide has tripled, from an average of 12 launches a year between 1990 and 1994, to between 36 and 37 commercial launches in each of the past three years. This is the result of increasing demand for geosynchronous communications satellites and the emergence of non-geosynchronous (NGSO), or low Earth orbit (LEO), communications constellations beginning in 1997. Several such systems—Iridium, Globalstar, and Orbcomm—were deployed between 1997 and 1999. As the number of commercial launches has rapidly increased, so too have forecasts for future launch activity.

Since mid-1999, however, a number of developments in the satellite communications market have led to a re-examination of those more optimistic forecasts. The most significant development was the bankruptcy of Iridium, the pioneering low Earth orbit mobile telephony constellation. As a result, many of the non-geosynchronous satellite constellations expected to be launched in the next five years are now facing increased skepticism from the financial community. These decreased expectations are reflected in the *2000 Commercial Space Transportation Forecasts*, published in May 2000 by the Federal Aviation Administration's Office of Commercial Space Transportation (FAA/AST) and the Commercial Space Transportation Advisory Committee (COMSTAC).

## 1. Introduction

While the number of commercial launches to non-geosynchronous orbits has rapidly increased over the past several years, the operators and proponents of non-geosynchronous orbit (NGSO) systems have suffered several significant setbacks over the past year. In particular, the pioneering Iridium Big LEO mobile telephony system which deployed 88 spacecraft on 20 launches failed to attract enough subscribers to continue operating and was compelled to file for bankruptcy protection. As a result, many of the non-geosynchronous satellite constellations expected to be launched in the next five years now face increased skepticism and appear less likely to be launched.

These decreased expectations are reflected in the soon-to-be-released *2000 Commercial Space Transportation Forecasts* [Reference 1], published jointly by the Federal Aviation Administration's Office of Commercial Space Transportation (FAA/AST) and the Commercial Space Transportation Advisory Committee (COMSTAC). The *2000 Commercial Space Transportation Forecasts* project only a modest increase above current levels of about 15 percent, down from the increase of 42 percent projected last year. While the increase is not expected to

11

*M. Rycroft (ed.), The Space Transportation Market: Evolution or Revolution?*, 11–17.

be as large as had been projected earlier [Reference 2], the FAA and COMSTAC nonetheless continue to expect demand for commercial space launches to grow above its present level.

## 2.    Recent Trends

### 2.1    Growth of Commercial Space Transportation

Over the past 10 years, the number of commercial space launches worldwide has tripled from an average of 12 launches a year between 1990 and 1994, to between 36 and 37 commercial launches in each of the past three years [Reference 3]. This dramatic growth is the result of increasing demand for geosynchronous (GSO) communications satellites and the emergence of non-geosynchronous (NGSO), or low Earth orbit (LEO), communications constellations (see Fig. 1). Several such systems consisting of multiple spacecraft in low Earth orbit—Iridium, Globalstar, and Orbcomm—were deployed between 1997 and 1999, opening an entire new market for commercial launches. While there were 18 launches to GSO in 1999, there were also 18 launches to LEO to deploy global satellite communications systems, remote sensing spacecraft, and developmental spacecraft.

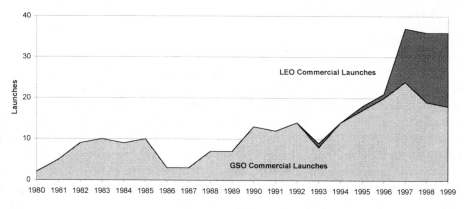

**Figure 1.** Historical commercial launches to GSO and LEO orbits

Annual commercial launch revenues have grown by two-thirds over the period from 1995 to 1999; 1995 revenues were about US $ 1.3 billion compared to US $ 2.2 billion in 1999. Revenues in 1999 were up only slightly from US $ 2.1 billion 1998, and are still lower than a high of US $ 2.4 billion in 1997 due to fewer GSO launches.

## 2.2   *Commercial versus Government Activities*

The rise in commercial launch activity has also steadily eroded the domination of space by government activities, with commercial activities now representing over 40 percent of worldwide launches conducted annually (see Fig. 2). As a proportion of total launches, the number of commercial launches continues to grow relative to government-sponsored launches—commercial launches represented 23 percent of total launches in 1995, 44 percent in 1998, and 46 percent in 1999. While commercial activity has greatly increased, government launches have dramatically declined as well. Since the end of the Cold War, worldwide government launches have fallen over 50 percent, from 106 in 1990 to only 43 in 1999, its lowest level since 1960. Most of this decline has been in launches by the former Soviet Union.

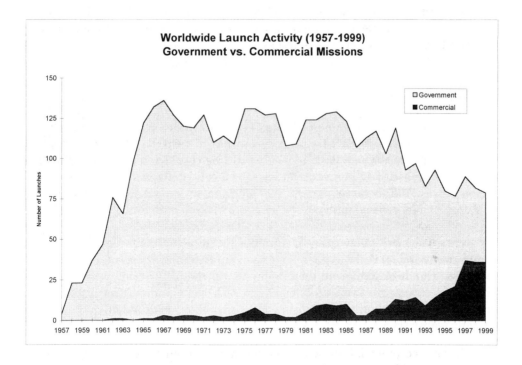

**Figure 2.** Worldwide launch activity (1957-1999), government vs. commercial

*2.3   Introduction of New Vehicles*

The industry also has diversified over the past five years, leading to increased competition for launches of commercial satellites. One of the most significant developments is the introduction of the Russian Proton and Soyuz launch vehicles and the international Sea Launch venture on the international launch services market. Each of these vehicles is now a major player in the satellite launch business, and is the result of a strategic partnership between a former Soviet launch vehicle manufacturer and a Western partner.

At the same time, longtime commercial launch providers Arianespace, Boeing, and Lockheed Martin have introduced, or are introducing, new vehicles capable of lifting heavier payloads. Ariane has now begun commercial operations of the Ariane 5 heavy-lift launch vehicle, designed to replace the Ariane 4. Boeing's Delta 3 is poised for its first successful launch following two failed launch attempts, and Lockheed Martin's Atlas 3 was successful on its first launch on 24 May 2000. And, in just a couple of years, the Delta 4 and Atlas 5 vehicles are expected to begin service after eight years of development under the U.S. Air Force's Evolved Expendable Launch Vehicle (EELV) program. Under EELV, the U.S. Air Force is investing US $ 1 billion, while Boeing and Lockheed Martin are investing over US $ 2 billion of their own funds.

## 3.   Future Activity

As the number of commercial launches has rapidly increased, so too have forecasts for future launch activity. For example, in 1999 the FAA projected that an average of 51 commercial launches would be conducted each year through 2010 — an increase of 42 percent over current levels [Reference 2]. This dramatic increase was due in large part to the plethora of proposals to deploy non-geosynchronous constellations.

Since mid-1999, however, a number of developments in the satellite communications market have led to a re-examination of those more optimistic forecasts. The most significant development was the bankruptcy of Iridium, the pioneering low Earth orbit mobile telephony constellation. After 20 launches deploying 88 spacecraft costing over US $ 5 billion, Iridium failed to attract enough subscribers to continue operating and was compelled to file for bankruptcy protection. As a result, many of the non-geosynchronous satellite constellations expected to be launched in the next five years are now facing increased skepticism from the financial community and appear less likely to be developed.

*3.1   2000 Commercial Space Transportation Forecasts*

The Federal Aviation Administration's Office of Commercial Space Transportation (FAA/AST) and the Commercial Space Transportation Advisory Committee (COMSTAC) have prepared projections of global demand for commercial space launch services for the period 2000 to 2010. [For a description of the FAA and COMSTAC, see section 4 below.] These projections are jointly published in *2000 Commercial Space Transportation Forecasts* [Reference 1]. This document includes:

- The *COMSTAC 2000 Commercial GSO Spacecraft Mission Model*, which projects demand for commercial satellites that operate in geosynchronous orbit (GSO) and the resulting launch demand to geosynchronous transfer orbit (GTO)

- The FAA's *2000 Commercial Space Transportation Projections for Non-Geosynchronous Orbits (NGSO)*, which projects commercial launch demand for all space systems in non-geosynchronous orbits, such as low Earth orbit (LEO), medium Earth orbit (MEO), and elliptical Earth orbits (ELI).

Together, the COMSTAC and FAA forecasts (see Fig. 3) project that an average of 41 commercial space launches worldwide will occur annually through 2010. This is an increase of 15 percent from the 36 commercial launches conducted worldwide in 1999. However, the forecast is down close to 20 percent from last year, which projected an average of 51 launches per year [Reference 2]. This downturn in expectations is the result of difficulties encountered by NGSO systems over the last year, including the failure of Iridium and bankruptcy of ICO.

Specifically, the forecasts shown in Table 1 [Reference 1] project that on average the following type and number of launches will be conducted each year:

- 23.5 launches of medium-to-heavy launch vehicles to GSO

- 7.5 launches of medium-to-heavy launch vehicles to LEO, or NGSO orbits

- 10.4 launches of small launch vehicles to LEO.

| | 2000 | 2001 | 2002 | 2003 | 2004 | 2005 | 2006 | 2007 | 2008 | 2009 | 2010 | Total | Avg |
|---|---|---|---|---|---|---|---|---|---|---|---|---|---|
| **Payload Demand** | | | | | | | | | | | | | |
| GSO | 30 | 31 | 35 | 31 | 32 | 31 | 30 | 28 | 30 | 29 | 30 | 337 | 30.6 |
| LEO | 23 | 19 | 29 | 74 | 62 | 59 | 74 | 45 | 37 | 56 | 74 | 552 | 50.2 |
| **Total Payloads** | **53** | **50** | **64** | **105** | **94** | **89** | **103** | **72** | **67** | **84** | **104** | **889** | **80.8** |
| **Launch Demand** | | | | | | | | | | | | | |
| GSO Med.-Heavy | 26 | 26 | 30 | 25 | 25 | 23 | 21 | 20 | 21 | 20 | 21 | 258 | 23.5 |
| LEO Med.-Heavy | 6 | 6 | 8 | 8 | 5 | 5 | 14 | 10 | 8 | 4 | 8 | 82 | 7.5 |
| LEO Small | 7 | 7 | 9 | 13 | 13 | 10 | 10 | 9 | 9 | 13 | 14 | 114 | 10.4 |
| **Total Launches** | **39** | **39** | **47** | **46** | **43** | **37** | **45** | **37** | **39** | **36** | **43** | **454** | **41.4** |

**Table 1.** 2000 commercial space transportation combined payload and launch projections [Reference 1]

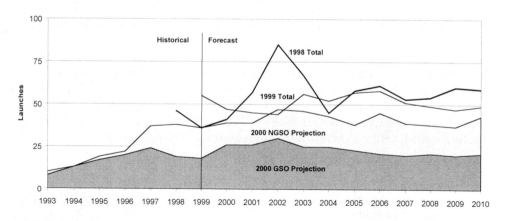

**Figure 3.** Combined GSO and LEO launch demand projections [Reference 1]

## 4.    FAA's Office of Commercial Space Transportation and COMSTAC

The Federal Aviation Administration's Associate Administrator for Commercial Space Transportation (FAA/AST) licenses and regulates U.S. commercial space launch activity as authorized by Executive Order 12465, Commercial Expendable Launch Vehicle Activities, and the Commercial Space Launch Act of 1984, as amended. AST's mission is to license and regulate commercial launch operations to ensure public health and safety and the safety of property, and to protect national security and foreign policy interests of the

United States during commercial launch operations. The Commercial Space Launch Act of 1984 and the 1996 National Space Policy also direct the Federal Aviation Administration to encourage, facilitate, and promote commercial launches.

The Commercial Space Transportation Advisory Committee (COMSTAC) provides information, advice, and recommendations to the Administrator of the Federal Aviation Administration within the Department of Transportation (DOT) on matters relating to the U.S. commercial space transportation industry. Established in 1985, COMSTAC is made up of senior executives from the U.S. commercial space transportation and satellite industries, space-related state government officials, and other space professionals.

The primary goals of COMSTAC are to:

- Evaluate economic, technological and institutional issues relating to the U.S. commercial space transportation industry

- Provide a forum for the discussion of issues involving the relationship between industry and government requirements

- Make recommendations to the Administrator on issues and approaches for Federal policies and programs regarding the industry.

Additional information concerning FAA/AST and COMSTAC can be found on the FAA's web site, at <http://ast.faa.gov>.

**References**
1.  U.S. Federal Aviation Administration and the Commercial Space Transportation Advisory Committee: *2000 Commercial Space Transportation Forecasts*, May, 2000
2.  U.S. Federal Aviation Administration and the Commercial Space Transportation Advisory Committee: *1999 Commercial Space Transportation Forecasts*, May, 1999
3.  U.S. Federal Aviation Administration: *Commercial Space Transportation: 1999 Year In Review*, January, 2000

# Future Market for Space Transportation
# EUTELSAT'S Perspectives

J. J. Dumesnil, S. Glynn, EUTELSAT, 70 rue Balard, 75502 Paris Cedex 15, France

e-mail: jjdumesnil@eutelsat.fr, sglynn@eutelsat.fr

**Abstract**

The space transportation market is close to a "redistribution of cards". All the major players are developing new heavy launch vehicles to meet the demand for larger satellites and to reduce the cost per kilogram launched. Larger, but not more numerous, GEO satellites will be launched in the future, whereas LEO constellations will not be the "eldorado" which was supposed to provide significant growth of the launch business. Therefore, it is likely that not all new launch vehicle developments will survive the competition. The winners will first be selected for their demonstrated reliability and performance, then for their price attractiveness and then for the quality of their services.

As a major operator, EUTELSAT will continue to launch satellites at a rate of more than one per year over the next decade. Apart from the major criteria mentioned above (reliability and cost), EUTELSAT will pay particular attention to the flexibility that it can be offered in terms of launch schedule and re-launch capabilities.

## 1. Introduction

Satellite operators cannot analyze and comment on all trends in the space transportation market with the same level of comprehensiveness as can the launch service providers. EUTELSAT is no exception: it has the knowledge of its own market segment, but cannot pretend to have any competence in the scientific satellite segment, which is anyway limited and heterogeneous, nor in the government satellite segment, which is almost non-existent in Europe but fairly important in the United States and in Russia. Therefore, EUTELSAT's considerations of the space transportation market will be focussed on the commercial segment, mainly communications and navigation satellites.

## 2. Commercial Satellite Evolution

The prime market segment over the past decade has been that of geostationary communications satellites, with launch masses ranging from 2 to 4 tonnes or even more. It is likely to remain the dominant segment in the years to come, with a continuing evolution towards higher masses. Clearly this evolution results from economic constraints imposed on satellite operators who are obliged to reduce their tariffs under the pressure of competition. Roughly speaking, historically, the satellite cost accounts for 40% of the cost per standard channel per year of operation, the launch for another 40%, insurance for 10%, the remaining 10% is shared between procurement monitoring, marketing and orbital operations. Reducing the satellite costs per channel-year means

19

*M. Rycroft (ed.), The Space Transportation Market: Evolution or Revolution?*, 19–23.
© 2000 *Kluwer Academic Publishers. Printed in the Netherlands.*

increasing the number of channels per satellite, and hence the development of new and bigger satellite platforms which can accommodate 50 channels or more and can operate for at least 15 years. Consequently, the satellite launch masses will soon generally exceed 4 tonnes and then reach 5 tonnes.

In parallel, the space transportation industry is developing more powerful launch vehicles capable of meeting the new satellite needs and, at the same time, reducing the cost per kilogram launched. For their business growth, much hope had been put by the launch service providers into another segment - LEO constellations. The Iridium and Globalstar satellites were indeed launched, but the economic uncertainties of such systems cooled the enthusiasm which prevailed a few years ago. A few other systems are under development in the telecommunications and navigation fields. They will allow one or two small- to medium-class launch vehicle operators to survive. Contrary to the predictions made some years ago, the frenetic commercialisation of de-commissioned missiles will not materialize.

## 3.    Launch Vehicles Evolution

During the past decade, the space transportation market was dominated by Ariane 4, the Atlas II family and, for lighter payloads, by Delta II. To be complete, one should add Proton and Soyuz when they became available internationally and commercially.

The race towards larger lift-off mass and cost reduction is having several effects. Ariane 4, Atlas II and Delta II will be phased-out gradually and replaced by new developments, respectively, Ariane 5, Atlas III and V, and Delta III and IV. Proton and Soyuz are being fitted with new, more powerful upper stages, Breeze-M and Fregat, respectively. New entrants are also attempting to take a share of the market — Sea Launch based on Zenit and the Japanese HIIA. To complete the picture for GEO satellites, one should add Long March 3A and 3B which currently suffer from political isolation.

As a result, the space transportation market is now in a critical period of transition between the old generation and the new. It becomes difficult to procure a launch on a vehicle of the old generation (or the lift-off performance is no longer adequate) and, on the other hand, very little is known of the intrinsic reliability of the new vehicles. Satellite operators must therefore select a launch vehicle which presents a higher degree of risk and which may not be available for launch when needed, for programmatic or for technical reasons.

This uncomfortable situation may last for several years, that is for the time needed to demonstrate a good probability of launch success and a reasonable control of operational availability. It will be very difficult to elaborate a business plan based on a long-term strategy on launch service procurement. Quantity buys cannot be considered. Instead, lack of visibility will force satellite operators to spread the risks and to maintain a quick reaction to events.

Considering that the commercial satellite launch market will not increase significantly in the next five years, at least in terms of numbers for the GEO satellite segment, and that there are now more launch vehicles competing on that market, it is unavoidable that there will be winners and losers. In the first instance, natural selection will apply, based on performance and reliability. Then the financial and operational conditions will come into play: price, payment plan, financing, insurance rates; re-launch/refund in case of launch failure, etc.. Finally the quality of services will be considered: mission integration, launch site facilities, launch campaign services, etc..

## 4.   EUTELSAT's Launch Policy

In the past, EUTELSAT has made exclusive use of Ariane 3 and 4 and Atlas II/IIA/IIAS to launch its first, second and third generations of satellites — in total, 17 launches up to the end of 1999, at which point in time 13 satellites were still operational (two EUTELSAT-I, four EUTELSAT-II, five Hot Bird and two W satellites). Fig. 1 shows the deployment as of January 2000 of these satellites together with Telecom 2A and DFS Kopernikus, which are commercially exploited by EUTELSAT.

Like every other satellite operator, EUTELSAT is now faced with the transition phase at the very moment when it needs to increase its launch rate. Its strategy clearly consists in spreading the risks. Year 2000 will be marked by 3 launches: SESAT on a Proton-Block DM, W4 on an Atlas IIIA and W1 on an Ariane 5. The first and third are due to provide continuity to existing services; therefore, their launches must be as safe as they possibly can be under the current circumstances. In contrast, W4 will open new services in the BSS bands over Russia and Africa, which explains why a slightly higher degree of risk could be contemplated in exchange for more favourable financial conditions for the use of the first Atlas IIIA launch.

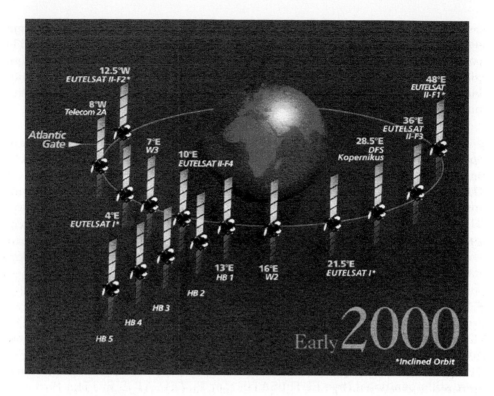

**Figure 1.** EUTELSAT's fleet as at early 2000

Next year, EUTELSAT will continue its diversified approach with two launches on Ariane 4 or 5 and a third one on Long March 3A. For its launches in 2002 onwards, EUTELSAT will firm up its choices as late as possible in order to integrate as many inputs as possible into its decision process. Typically a decision 12 months prior to the desired launch date should be acceptable.

If all the upcoming launches are successful and with the decommissioning of old satellites, the EUTELSAT fleet will consist of 19 operational satellites as of early 2002. The EUTELSAT fleet of satellites is typical of the launch mass evolution imposed on the satellite operators by the competition constraints; one tonne for the first generation, two tonnes for the second one and now 3 tonnes for the third one, with a peak of 3.8 tonnes for HOT BIRD™ 6.

Current studies forecast that EUTELSAT's next generation of satellites will range between 4.5 and 5 tonnes in the 2003-2005 time period. On paper, six or seven types of launch vehicles would be capable of doing the job, provided of course that they have demonstrated their reliability through several consecutive successful launches. It is worth noting that the reliability of a given launch vehicle will translate directly into a launch insurance rate that the underwriting community will be willing to quote for that launch vehicle.

The next selection criteria will be of financial nature. Launch service providers will have to be able to propose attractive price conditions for an operational launch (i.e. not a qualification or demonstration launch), most probably not higher than those currently prevailing for a 3 to 3.5 tonne satellite. EUTELSAT expects to have included in the price all services and facilities that are to be provided on the launch site (this is not always the case), and also a free re-launch in case of launch failure, or alternatively a refund if no replacement satellite is available or foreseen.

Availability/flexibility is another important aspect of EUTELSAT's launch policy. Satellites must be launched as soon as they are ready. To that effect, EUTELSAT has been able in the past to decide between its various satellites on order and its launch contracts, and it wishes to continue to do so in the future. This means that the satellite manufacturers must be able to maintain compatibility of their satellites with a wide variety of launch vehicles, including through the vibration tests. Conversely, the launch service providers must be prepared to accept a swap of satellites at a late stage in the programme.

A discussion of launch service requirements would not be complete without a section on the technical visibility. For themselves as well as for their obligations to the insurance underwriters, satellite operators must be granted full technical visibility of the launch vehicle manufacture, integration and test. Every failure or anomaly must be recorded and reported. This was a presumption accepted by everyone until March 1999 when the US Government decided to enforce a tighter control on export licenses and data transfer. This is certainly prejudicial to the US launch service providers. But, equally, EUTELSAT cannot and will not enter into contract with a company that has not obtained the necessary approvals beforehand.

# The Coming Commercial Passenger Space Transportation Market

**P. Collins,** Space Transportation Planning Department, National Space Development Agency, 2-4-1 Hamamatsu-cho, Minato-ku, Tokyo 105-8060, Japan

e-mail: Patrick.Collins@nasda.go.jp

### Abstract
This paper discusses the future of commercial passenger space transportation from multiple viewpoints - technical, financial, organisational and political - as well as economic policy, legal and social considerations. It argues that passenger space travel offers the opportunity for launch services to grow into a profitable commercial industry, with potential to grow to a very large scale, like passenger air travel. However, addressing this new market requires collaboration between space and aviation interests – which the space industry has resisted to date. At a time of record levels of unemployment in many countries both rich and poor, it is both economically and socially desirable that this resistance is overcome and this popular new service be developed as soon as possible.

## 1. Introduction

Markets for transportation services are driven by customer demand: unless the demand for a new capability or service is forecast to be sufficiently large, the cost and risk of innovating will not be undertaken by investors. Space transportation follows the same rules: the market for satellite launches, which has a turnover of just a few US $ billion/year, has supported the derivation of expendable launch vehicles (ELVs) from missiles developed for defence purposes, but none is profitable on a fully costed basis. Although it is well-known that the development of reusable launch vehicles (RLVs) could reduce the cost of launch, in the absence of major new demand for launch services it would not be profitable to develop them (unless their development cost could be kept very low - below US $ 1 billion).

Recently a huge new potential market for launch services has been identified — the demand for commercial passenger space travel or 'space tourism'. Market research in the leading countries — USA, Japan, Germany, Britain, Canada — shows that there is a huge pent-up demand for space travel services, even at relatively high prices. Furthermore, no other large market for launch services has been identified. Consequently, until the passenger market is tackled, launch costs will remain high and human space activities will remain largely a burden on taxpayers. However, passenger space travel is not yet receiving serious attention from government space agencies, which have other interests.

*M. Rycroft (ed.), The Space Transportation Market: Evolution or Revolution?*, 25–34.

## 2.    Technical Considerations

The technical issues relating to the feasibility of passenger space travel can be divided into three parts:

### 2.1   Sub-orbital Flights

Whereas flying to orbit requires acceleration to a velocity of some 8 km/s, flying to 100 km altitude requires a velocity of only 1 km/s. As a consequence, vehicles capable of providing sub-orbital space flights are currently under development by a number of companies, although raising finance in the current climate is very difficult. It is notable that a well-designed approach has no technical risk, since all the necessary components and know-how have long been available. Travel companies have customers waiting to take such a trip, even at prices as high as US $ 100,000/passenger, although these are expected to fall eventually to less than US $ 10,000/passenger.

### 2.2   Orbital Flights

The next technical development will be vehicles capable of making passenger flights to low Earth orbit, building on the accumulated experience of sub-orbital operations, and following a path of incremental testing as used in commercial aviation. Both single-stage-to-orbit, vertical take-off and landing (SSTO VTOL) and two-stage-to-orbit, horizontal take-off and landing (TSTO HTOL) vehicles are technically feasible with known technology. The main challenge is to design and develop systems with the optimal combination of operational lifetime, ease of maintenance, operating costs, turnaround time, passenger capacity, production costs, safety and comfort.

For both VTOL and HTOL vehicles, while many components will be taken from or based on those used in existing launch vehicles, the overall design process will be closer to that of passenger aircraft design: reusable launch vehicles' specifications will be decided by prospective operating companies which will be the customers for the vehicles' manufacturers. Consequently airlines' and aircraft manufacturers' experience is more relevant than experience of building and launching expendable satellite launch vehicles.

In addition, passenger-carrying space vehicles will pass through the atmosphere at the start and end of orbital flights, during which they will be subject to aviation regulations. Thus, for example, it seems probable that Vertical-Take-Off-Horizontal-Landing (VTOHL) vehicles will not be used for passenger-carrying, due to the difficulty of their meeting the basic aviation requirement of savability — namely that the vehicle and its passengers should be able to survive any credible accident at any time of flight from take-off to landing. Since non-passenger-carrying vehicles will not be profitable, and will therefore be a continuing burden to taxpayers, publicly-funded work on reusable launch vehicles should be restricted to passenger-carrying vehicles. It is desirable that work should be done on both SSTO VTOL and TSTO HTOL approaches, in a competitive and international framework that will be able to evolve naturally into a commercial activity.

### 2.3   Orbital Accommodation

The third technical development will be facilities providing accommodation in orbit for guests wishing to stay for several days. Technically, orbital accommodation vehicles are much simpler than passenger launch-vehicles, and indeed simpler than orbital science research stations, since they require little specialised equipment. Test flights of reusable launch vehicles may be used to launch components of accommodation facilities even before orbital passenger flight services begin. Accommodation is the only potentially large market in orbit for habitable modules, such as those being developed for the International Space Station.  As in normal businesses, having spent billions of dollars developing these modules and their components, manufacturers would like to find customers to purchase tens or even hundreds of them. Several large companies are starting to perform design work on orbital accommodation facilities, and MirCorp is already selling orbital accommodation services on a commercial basis.

### 3.   Financial Considerations

In order to earn a commercial rate of return on an investment of several US $ billion, it is necessary to earn profits of the order of US $ 1 billion/year. In line with this, executives of Lockheed Martin Corporation have confirmed that the proposed "Venture Star" reusable satellite launch vehicle cannot be developed commercially because the market for launching satellites and government payloads is too small to repay the estimated investment of US $ 6 billion. By contrast, passenger-carrying vehicles could repay the investment needed for their development, due to the far larger potential market [Reference 1].

A single RLV that could launch even once/week could do all the work of expendable satellite launch vehicles — and so is an unattractive investment for ELV makers. However, passenger travel offers a launch market that could grow hundreds of times larger than satellite launch. In addition, just as the turnover and employment generated by airline companies is many times larger than that of aircraft manufacturers, so RLVs will create many times more business and employment than their manufacture alone.

According to fundamental economic reasoning, the value of an activity is the (discounted) present value of the profits to which it will lead in the future. Based on market research-based estimates of the potential market for space tourism [Reference 2], and on expert estimates of the cost of developing passenger launch vehicles [Reference 1], the discounted present value of a space tourism industry development scenario is strongly positive [Reference 3] — indicating that the sooner that it is developed the greater the economic benefit. Bekey [Reference 1] estimates that annual profits of US $ 30 billion could eventually be earned from an initial investment of just US $ 7.6 billion.

It is important to recognise that it is not necessary for space tourism services to be profitable from their inception in order for it to be economically desirable to develop them. This is because civilian government space activities, for which taxpayers currently pay some US $ 25 billion/year, are not profitable by normal commercial standards: if they were profitable they would generate new commercial revenues growing by some US $ 25 billion/year/year, and new employment growing by around 1 million people/year/year. Instead, to date they have shown a hugely negative rate of return. From the economic point of view, a **partially** loss-making activity would be preferable. Thus, for example, if the operating costs of a first passenger-carrying orbital vehicle such as the Japanese Rocket Society's 'Kankoh-maru' were even 10 times higher than the target price of approximately US $ 1 million/flight, it could still be an economically valuable development, and become the basis for a Mark 2 vehicle with lower costs.

## 4.    Organizational Considerations

The US government's Office of Commercial Space Transportation (OCST) was moved into the Federal Aviation Administration (FAA) in 1995. The OCST is actively promoting the development of space transportation into "...a real mode of transportation... when a multitude of entrepreneurs will open space to all kinds of activities: thrill-rides, vacationers, industry and even trips to the Moon and beyond" [Reference 4], and its budget is being doubled.

This is desirable as the contemporary aviation industry provides a complete organizational model of the coming passenger space transportation industry. With almost a century of experience of developing and operating advanced technology transportation systems, a current scale of more than 1 billion passengers/year, and a global network of national and international services and regulations, the extension to include flights to and from space is but an incremental step. Indeed, the extension of air traffic control to include low Earth orbits and the licensing of commercial vehicles returning from orbit to Earth are already being studied within the FAA [References 5, 6].

Due to the wide accumulated experience of the aviation industry — including decades of experience of operating reusable, liquid-fueled rocket-engines on passenger-carrying vehicles — the space industry should establish mechanisms to collaborate actively with aviation organisations to make progress towards passenger space travel [Reference 7]. Joint research should be carried out on a wide range of subjects, illustrated in the following list:

## Life Sciences
Short-term orbital stays by average people — not long-term stays by
  statistically unusual people
Tests of the full range of anti-motion sickness medications
Effects of alcohol consumption in micro-gravity
Treatment of colds, nasal congestion, influenza and other common ailments
  in micro-gravity.

## Propulsion
Design for Ease of Maintenance (DEM, widely applied in other industrial
  sectors) applied to rocket engines and vehicles
Accumulation of operating data on rocket engine reuse (including study of
  historical Rocket-Assisted-Take-Off data)
Rocket engine reliability increase
Rocket engine noise suppression
Cost reduction rather than mass reduction.

## Transportation
Study of optimal sizes of passenger space vehicles
Passenger vehicle optimisation for specified routes/services
Orbital propellant storage / in-space photo-voltaic electrolysis of water / in-
  space cryogenic liquid transfer.

## Orbital Accommodation
Large windows
Micro-gravity plumbing
Facilities for large populations (more than 100 guests)
Cost reduction through mass production
Rotating accommodation.

**Legal issues**
  Extension of aviation regulations to cover space flight
  Jurisdiction in orbit
  Introduction of liability for damage in space, including that caused by
      space debris
  Promulgation of space salvage law based on international marine law.

**Insurance**
  Early contribution to vehicle design
  Standardisation of procedures, such as developing 'AVN 67B of Space'
      [Reference 8]
  Building standards.

## 5.    Political Considerations

The budgets of government space agencies depend on popular support, albeit indirectly. Market research has shown that space tourism is an extremely popular idea with the general public, so there is little doubt that, if given the choice, the public would support the use of public funds to accelerate its realisation. However, space agencies have developed a political constituency which has to date defended their funding in most countries, and they are reluctant to tackle a subject as innovative as passenger travel for which they have little relevant experience, despite its economic value and general popularity.

In the past, governments have made major investments in successive generations of passenger transportation systems — canals, railways, highways, aviation — and it is desirable that they do so again for passenger space transportation, for the same reasons: to stimulate the growth of new commercial activities. However, space agencies have shown strong resistance to helping to realise this activity, despite the fact that it is greatly in the economic interest of the general public. Currently not 1% of their funding is used for research relevant to passenger space travel, and they have never surveyed the public to determine the level of support for the idea. As a result, it is now being proposed that responsibility and corresponding funding to develop passenger space travel should be given to other bodies than space agencies [Reference 9]. If this happens, space agencies may lose much of their remaining public support.

While it should not be government's role to operate space tourism services which, like other travel services, would be better operated by private companies, there are nevertheless many activities that governments should perform to facilitate the early development of a commercial space tourism industry.   Many of these are listed and discussed in the NASA/STA report 'General Public Space Travel and Tourism' [Reference 10], including:

"The Federal government... should inform the general public about space travel and tourism possibilities; such communications should focus on the idea that ordinary people — not just astronauts — should be able to go on a space trip in the relatively near future as a result of government-private sector cooperation".

The policy decision by the current NASA leadership not to follow this advice — and indeed not even to make the report accessible to the public — imposes major economic costs on taxpayers in the USA and elsewhere by delaying the development of a profitable commercial space travel industry. Government space policies should be revised to match the post-cold-war economic realities that are driving sweeping change in other industries [Reference 11].

## 6.    Economic Policy Considerations

If developed over the coming decade, a space tourism industry could grow to have a turnover in excess of US $ 100 billion by 2030 [References 1, 3]. This would be economically beneficial to the many companies and people directly and indirectly involved in the activity. In addition, such a development will also have important global and macro-economic benefits, due to the deflationary pressures in the world economy today, which are similar in some respects to those experienced during the 1930s, being also preceded by a stock market 'bubble' in the USA [Reference 12].

Economic growth is driven by companies continually improving their efficiency by a few % per year; progressive growth of unemployment is avoided by the creation of new industries which employ those displaced from older industries, as seen in the long list of new industries that grew into major employers through the 20th century. The major changes in the world economy in recent decades — notably the collapse of the Soviet Union which released several hundred million people to join the world market, the rapid growth of countries in South East Asia to where a large amount of manufacturing industry has moved due to lower costs, and the rapid progress in computers and communications that has reduced the cost of obtaining and handling information — have all increased competition in existing industries, leading to some 20% excess capacity in mature industries such as motor-manufacturing, steel and petro-chemicals [Reference 11].

The passenger space travel industry has the economic potential to grow through the 21st century as aviation did during the 20th century, creating new employment on a very large scale and helping to counteract current deflationary pressures. This possibility is much too important to be allowed to continue to stagnate as it has for the past 20 years due to government-funded space agencies' reluctance to innovate.

## 7.    Legal Considerations

The growth of commercial aviation has led to the development of a seamless global network of legal arrangements, including international treaties, involving manufacturers, operators, financiers, insurers, governments and the general public. Space law consists of a number of inter-governmental treaties some aspects of which are inappropriate for passenger space travel. The meshing of these two fields of law is a most important and challenging task for legal experts — and government funding should be made available for it. As an example, the FAA's proposals to extend Air Traffic Management systems to include low Earth orbit, and to establish an International Space Flight Organisation (ISFO) [Reference 5], are initiatives that should already be the subject of active debates around the world [Reference 13] — but to date no other government has responded at all. In the face of such willful lack of foresight among other governments, if US domination of this field were to become a *fait accompli*, the US government certainly could not be criticised.

## 8.    Social Considerations

Last but not least, the development of a vigorous space tourism industry will have important social benefits. From historical precedents it is known that the opening of a major new geographical region for commercial development has an economically stimulating effect which creates optimism pervading society. It is hard to be specific about such benefits, but they will be especially notable for the younger generation in the richer countries, who today suffer problems of lack of motivation and self-discipline due to a lack of real challenges, and due to a mistaken but widespread pessimism that humankind's future is limited.

In truth, humans stand on the brink of a major expansion of economic growth into space. However, it can be motivated and financed only by the development of commercial passenger space travel services. This will provide a uniquely fascinating and constructive new focus for the energy and creativity of the younger generation around the world. To continue to deny them this opportunity due to the resistance to change by the vested interests of government space agencies and their clients would be scandalous.

## 9.  Conclusions

In view of the technical feasibility and the economic promise of passenger space travel, as well as the lack of any other direction for significant growth of commercial space activities, the mobilisation of concerted efforts to realize passenger space travel services in the near future is highly desirable. One way to start is for governments to follow the excellent advice published in the NASA/STA report on General Public Space Travel and Tourism, including that quoted above [Reference 10].

In order to make substantive progress in this direction, budgets should be made available for relevant research in a wide range of associated fields, particularly involving collaboration between aviation and space organisations. If government space agencies will not embrace this new challenge, it will be economically very beneficial if their funding is reallocated to suitable organisations, primarily aviation-related, which will do so [Reference 3].

### References

1.    Bekey, I: *Economically Viable Public Space Travel*, Proceedings of 49th International Astronautical Federation Congress, 1998, also downloadable from <www.spacefuture.com/archive/economically_viable_public_space_travel.shtml >.
2.    Collins, P et al: Demand for Space Tourism in America and Japan, and its Implications for Future Space Activities, Advances in the Astronautical Sciences, Vol. 91, pp. 601-610, 1996, also downloadable from <www.spacefuture.com/ archive/demand_for_space_tourism_in_america_and_japan.shtml>.
3.    Collins, P: *The Space Tourism Industry in 2030*, Proceedings of Space 2000, American Society of Civil Engineers, pp. 594-603, 2000, also downloadable from <www.spacefuture.com/archive/the_space_tourism_industry_in_2030.shtml>.
4.    Smith, P: Reliability and Space Transportation, *Space News*, Vol. 10, No. 30, p. 15, 1999.
5.    Smith, P: *Concept of Operations for the National Airspace 2005*, Federal Aviation Administration, 1999, also downloadable from <www.spacefuture.com/archive/concept_of_operations_in_the_national_airspac e_system_in_2005.shtml>.
6.    Anon, DRAFT Interim Safety Guidance for Reusable Launch Vehicles, Federal Aviation Administration, 1999
7.    Collins, P & Funatsu, Y: *Collaboration with Aviation - The Key to Commercialisation of Space Activities*, Proceedings of 50th International Astronautical Federation Congress, 1999, also downloadable from: <www.spacefuture.com/archive/collaboration_with_aviation_the_key_to_comme rcialisation_of_space_activities.shtml>.

8.   Margo, R: Aspects of Insurance in Aviation Finance, *Journal of Air Law and Commerce*, Vol. 62, pp. 423-477, 1996

9.   Rogers, T: *Presentation to Year 2000 FAA Space Transportation Conference*, Federal Aviation Administration, Arlington, VA, February 8, 2000

10.  O'Neil, D et al: *General Public Space Travel and Tourism – Vol. 1, Executive Summary*, NP-1998-03-11- MSFC, National Aeronautics and Space Administration/Space Transportation Association, 1998, also downloadable from: <www.spacefuture.com/archive/general_public_space_travel_and_tourism.shtml>.

11.  Collins, P: *Space Activities, Space Tourism and Economic Growth*, Proceedings of 2nd International Symposium on Space Travel, Bremen, 1999, also downloadable from: <www.spacefuture.com/archive/space_activities_space_tourism_and_economic_growth.shtml>.

12.  Kuttner, R: Yes, Virginia, There Is a Speculative Bubble, *Business Week*, No. 3661-991, p. 12, April 17, 2000

13.  Collins, P & Yonemoto, K: *Legal and Regulatory Issues for Passenger Space Travel*, Proceedings of 49th International Astronautical Federation Congress, 1998, also downloadable from: <www.spacefuture.com/archive/legal_and_regulatory_issues_for_passenger_space_travel.shtml>.

# Business Opportunities for Space Transportation Companies in the Evolution of a Global Society

M. Bosch, Business Administration and Business Informatics, University of Applied Sciences Albstadt-Sigmaringen, 72488 Sigmaringen, Germany

e-mail: bosch@fh-albsig.de

### Abstract

Investments in space transportation companies (developers, providers, operators) are long-term undertakings. Therefore, the respective business plans have to take into account the long-term trends predicted by market research. According to the experts of leading think tanks, globalization is the most important fundamental trend. In answer to the question, "which trends will have the most important influence in the future?", 83 percent of the leading German think tanks named either "globalization" or "world-wide networks". The trend towards globalization is irreversible. This, in turn, will lead to a global society, which can be characterized as an intelligent, cybernetic system comparable to a living organism. The evolution of this system is characterized by the following attributes: (1) increasing world-wide interconnection by growing telecommunications networks, (2) rapid acceleration of economic processes, (3) increasing specialization of persons, enterprises, nations and economic regions, and (4) increasing interdependencies between specialized subsystems.

In this paper, the business opportunities in the space transportation market are derived from the characteristics (1) – (4) mentioned above. These opportunities are quantified, as far as possible. The paper starts out by describing the phenomenon of globalization. Next, the effects of the growing satellite-based telecommunications networks on the space transportation market are analyzed. After that, the relationships between world-wide networks, the acceleration of economic processes and the possibility to create a market for hypersonic business travel and transportation are presented. Due to the increasing specialization and the global interdependencies which result from it, the necessity for a comprehensive, "sensory", satellite-based monitoring and reconnaissance system is discussed. The effects of this system on the space transportation market are examined. Next, the consequences of the expanded use of satellites and of increasing hypersonic travel and transportation for Earth-based astronomy and space research are described. Finally, a model of the future development of the space transportation market summarizes all points of discussion.

## 1.   Introduction

A clear definition of the business strategy is one of the most crucial aspects to any business plan. Before the business strategy can be established, several prerequisites are necessary. First, the enterprise goals and core business areas need to be determined. Next, the relevant market must be defined and then predictions about the future development of this market must be extrapolated. The development of an enterprise strategy consists of a description of the steps required to achieve these enterprise goals by pursuing the core business areas in the relevant markets.

Due to the long and costly research and development phases in the space transportation industry, the establishment and enlargement of an enterprise in

35

*M. Rycroft (ed.), The Space Transportation Market: Evolution or Revolution?, 35–42.*

the space transportation market must be seen as a long-term undertaking. The enterprise strategy of a space transportation company must therefore be based on long-term market projections. Long-term trends need to be taken into account. According to the experts of leading think tanks, globalization is the most important fundamental trend. In answer to the question, "which trends will have the most important influence in the future?", 83 percent of the leading German think tanks named either "globalization" or "world-wide networks" [Reference 1].

The growth of world-wide networks based on the Internet and other types of telecommunications has resulted in the birth of an intelligent global system. This system surpasses all of the long-term efforts in computer science to develop artificially intelligent systems. If each of the individual participants or computers connected to the world-wide communications network were interpreted as synapses and their connections were defined as axons, then a level of connectivity comparable to the human brain will be reached in a few years. For example, a single participant is connected to the world-wide telecommunications network by telephone, cellular phone, fax, the World Wide Web, e-mail, Short Message Service (SMS), television and radio. A further analogy to a living organism is inherent in the increased specialization of business actors, enterprises and other subsystems and their computer-based interconnections to the global economy, such as outsourcing and business-to-business platforms on the Internet.

A living organism also has highly specialized subsystems which cooperate and are controlled by the central nervous system. The overall goal is the establishment and the maintenance of the system itself. According to systems theory, the global economy is thus a self-organizing, cybernetic system.

The space transportation business plays an important role in building and controlling this global system. As illustrated in this paper, space transportation is an important bottleneck in the further development of a global society. This bottleneck contains a reservoir of unfulfilled potential demand.

## 2.    Globalization and the Space Transportation Market

Globalization is not a steady state, but rather an evolutionary, cybernetic process. Therefore, the main "goal" of globalization is the development of an intelligent, self-controlled global system. In Fig. 1, the interconnections between globalization and the development of the space transportation market are shown graphically. An increasing level is marked with (+), a decreasing level with (-).

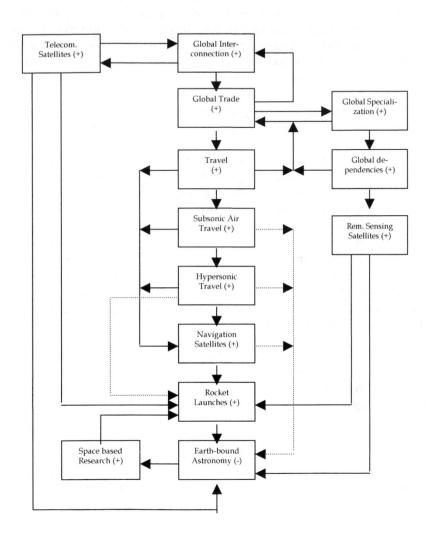

**Figure 1.** Globalization and the space transportation market

## 3.    Market Drivers and Restraints

### 3.1   Satellite Communications

An important prerequisite for world-wide networks is an efficient telecommunications infrastructure. As described above, telecommunications networks function as the central nervous system for the global organism. Satellites are the most efficient means for transferring data. As networking increases, the demand for telecommunication satellites will also increase. At the same time, an increase in data transfer volume leads to an increase in world-wide networking. The result is a positive feedback system.

In the next few years, it is to be expected that a large portion of the classical economy will be transformed into a new, Internet economy. Examples of this trend can be seen in the rapid development in the areas of e-commerce, business to business (B-2-B) solutions and mobile communications. Data transfer is a strategic bottleneck which holds back further growth in this area.

Broadband satellite services play an important role in the transfer of multimedia Internet data. Frost & Sullivan forecasts that the broadband satellite service market will grow from $ 350 million in 2000 to $ 9 billion in 2004 [Reference 2]. But this high growth rate stems in part from the migration of traditional satellite services from legacy satellites to newer, broadband ones. Furthermore, satellite operators will be able to tap demand in areas that are served by high-speed landline systems if they can remain price-competitive with those services. Widespread takeup of data-intensive applications, such as desktop videoconferencing or high-speed file transfer, will drive demand for all broadband communications technologies including satellites [Reference 2].

The Iridium disaster, the reorganization of ICO-Global under Chapter 11 bankruptcy code and the poor development of Globalstar has let to disillusionment in the satellite phone market. On the other hand, there is still a demand for satellite- based phone services in regions without terrestrial phone lines. In each of those regions, satellite telephones can provide immediate access to the rest of the world through traditional phone lines and the Internet. Building the same infrastructure on the ground will take decades and cost far more than space-based systems. A reduction in launch costs would definitely help to increase competitiveness for these systems [Reference 3].

### 3.2   Air/Space Travel and Transportation

In spite of the coming age of electronic communication and video conferencing, the increase in the global exchange of information, goods and services will lead to an increase in travel activity and in air travel [Reference 4]. Unpleasant delays prove that in many regions the maximum capacity for air travel has already been reached. The projected growth rates in air travel will lead to traffic jams similar to those on the streets today. In the new economy, where speed is an important competitive factor, punctuality and the shortest possible travel times are especially important for business people. Delays on land or in the air and long travel times with subsonic aircraft are no longer acceptable. A demand for super/hypersonic travel and transportation services, and an increased interest in space tourism, could have a positive effect on the emerging market for rocket planes.

One effect of the growth in travel which is already evident today is the increased demand for satellite-based navigation services. Other applications of those systems are fleet management, asset tracking, precision agriculture, time stamping for telecommunications, banking transactions and integration into precision weapons. In addition to the existing U.S. "GPS" and the Russian "GLONASS" systems, the EU plans its own satellite navigation system named "Galileo", which should be ready for use before 2010. GPS and GLONASS have upgraded their systems by increasing the number of satellites available to the user [Reference 5].

## 3.3   Remote Sensing

The growth in the global exchange of information, goods and services has led to an increase in specialization. Enterprises concentrate on their strategic core business areas and outsource all activities which are not part of these core areas. Thus, the international division of labor leads to an increase in specialization for people, enterprises and countries. Specialization has become one of the most important requirements for economic success.

Specialization also leads to world-wide interdependencies between business subjects. Thus, the entire global economy can be described as a highly networked, cybernetic system. Disturbances, such as war, terrorism, economic crises, raw material shortages, natural and environmental catastrophes, which only have an effect on subsystems, can still trigger a chain reaction crisis in the entire global economy. In general, increased specialization always results in an increased interdependence between specialized subsystems. Both specialization as well as the resulting interdependencies require an increase in the global exchange of information, goods and services. The interconnections shown here represent a positive feedback system.

In order either to avoid the effects of disturbances to the system or to detect the disturbances early enough to react, the global system needs a sensory organ in the form of commercial, scientific and military remote sensing and reconnaissance satellites. These systems enhance the accuracy of measurements and lead to a greater efficiency in mapping, environmental monitoring, arms treaty compliance, mineral exploration and agriculture. The demand for such satellites increases as global interdependencies increase. This leads directly to an increase in the demand for rocket launches. Furthermore, new sensing technologies such as high-resolution systems, synthetic aperture radar (SAR) systems or hyperspectral remote sensing systems are replacing conventional technologies. Thus, they generate an additional demand for launching services in the space transportation market. Replacement systems for non-functional satellites also increase the demand for launching services. Both high launch costs and regulatory issues limit growth in this area.

### 3.4   Space-Based Astronomy and Human Spaceflight

Earth-based astronomy will be even more limited by the increase in communications, navigation and observation satellites and by space debris in LEO (low Earth orbit) and MEO (medium Earth orbit). The increasing number of rocket launches, the increase in air and space travel, as well as the resulting increase in air pollution will decrease the quality of Earth-based astronomical observations. As a result, space research will have to be accomplished with the help of space telescopes, probes, and manned space missions, such as a manned mission to Mars. An observation station on the Moon would be an optimal base for astronomical explorations. Each of the possibilities listed here would once again lead to an increase in the demand for rocket launches. Limiting factors here are primarily the high costs of launching, the limited government budgets for science and the amount of funding provided by private sponsors for space exploration.

### 4.   Computer-Based Simulation with System Dynamics

Computer-based simulation with System Dynamics has proven to be an effective tool for modeling non-linear, asynchronous and complex systems, where traditional operations research methods fail. A System Dynamics model consists of a time loop, which represents the individual time periods. In each period, the levels are varied according to the inflow and outflow rates [Reference 6]. The following model in Pascal pseudo code  is based on Fig. 1. It contains a total of 13 levels with their respective inflow and outflow rates. These levels are updated in each time period. The implementation of this model with explicit parameters can be used to predict the dynamic development of the space transportation market.

```
PROGRAM system_dynamics;

VAR (* Declaration of variables *)
  x₁,
  .
  .
  .
  xₙ: REAL; (* levels x₁ to xₙ *)
  simulated_period,
  defined_simulation_end: INTEGER;
  .
  .
  .

PROCEDURE initialization;
BEGIN
  (* Systems initialization *)
END; (* of initialization *)

PROCEDURE set_input_parameters;
BEGIN
  (* This procedure allows the user to enter initial values and other
     input parameters *)
END; (* of set_input_parameters *)

FUNCTION dx₁_increase: REAL;
BEGIN
  dx₁_increase := f(interaction of levels increasing x₁);
END; (* of dx₁_increase *)

FUNCTION dx₁_decrease: REAL;
BEGIN
  dx₁_decrease := f(interaction of levels decreasing x₁);
END; (* of dx₁_decrease *)
  .
  .
  .
FUNCTION dxₙ_increase: REAL;
BEGIN
  dxₙ_increase := f(interaction of levels increasing xₙ);
END; (* of dxₙ_increase *)

FUNCTION dxₙ_decrease: REAL;
BEGIN
  dxₙ_decrease := f(interaction of levels decreasing xₙ);
END; (* of dxₙ_decrease *)

PROCEDURE update_levels;
BEGIN
  x₁ := x₁ + dx₁_increase - dx₁_decrease;
  .
  .
  .
  xₙ := xₙ + dxₙ_increase - dxₙ_decrease;
END; (* of update_levels *)
```

```
PROCEDURE save_time_step_values;
BEGIN
  (* This procedure saves the current period values of levels and rates
     on the harddisk for statistical purposes *)
END; (* of save_time_step_values *)

BEGIN (* of main_program *)
  initialization;
  set_input_parameters;
  REPEAT
    INC (simulation_period);
    update_levels;
    save_time_step_values;
  UNTIL simulation_period = defined_simulation_end;
END. (* of main_program *)
```

## 5.    Conclusions

This paper has shown that the space transportation market is a strategic bottleneck for the development of a global economy and of a truly global society. The interactions discussed here are, for the most part, qualitative rather than quantitative in nature. In order to achieve a useful prognosis, however, quantitative parameters for the simulation model are necessary. Future research in this area will concentrate on developing these quantitative parameters. In conclusion, it would be helpful to post a prototype of the simulation model on the Internet, so that experts worldwide would be able to work on its further development, to define quantitative parameters and to improve the quality of the data.

**Acknowledgements**
I would like to give special thanks to Patricia Shiroma Brockmann for her help in the translation of this paper into English as well as for her continued encouragement and thought provoking discussions.

**References**
1.    Own calculations based on manager magazin, 12/1999
2.    Frost and Sullivan: Introduction in Directory of Worldwide Space 2000, pp. 25 – 26, 2000
3.    Commentary: Iridium Is Not the End, in: *Space News*, Vol. 11, No. 12, p. 14, 2000
4.    Zubrin, R.: Entering Space, p. 52, 2000
5.    Frost and Sullivan: Introduction in Directory of Worldwide Space 2000, Space News, p. 8, 2000
6.    Niemeyer, G.: *Kybernetische System- und Modelltheorie*, pp. 218–220, Verlag Franz Vahlen GmbH, München, Germany, 1977

# Report on Panel Discussion 1:

# What will Constitute the Future Market for Space Transportation Services? Users' Perspectives

R. Leon, V. Tessier, International Space University, Strasbourg Central Campus, Parc d'Innovation, Boulevard Gonthier d'Andernach, 67400 Illkirch-Graffenstaden, France

e-mail: leon@isu.isunet.edu, tessier@isu.isunet.edu

**Panel Chair : P. Smith, FAA Office of Commercial Space Transportation, USA**

**Panel Members:**

**K. Bahrami,** MSS student representative, International Space University, France
**M. Bosch,** University of Applied Sciences Albstadt-Sigmaringen, Germany
**P. Collins,** NASDA, Japan
**J. Friis,** Final Analysis INC., USA
**S. Glynn,** EUTELSAT, France
**R. Khadem,** INMARSAT, UK
**B. Madauss,** Société Européene des Satellites S.A., Luxembourg
**J. Schafer,** NASA Headquarters, USA
**P. Smith,** FAA Office of Commercial Space Transportation, USA
**R. Tumlinson,** Space Frontier Foundation, USA

The discussion, which was rather more of a question-answer period, focused on the user's perspective of the future of space transportation; members of the panel gave their points of view on the topic.

**R. Tumlinson** mentioned that Space Frontier Foundation's main interest is to ensure reliable and safe access to space and to identify destinations in space. They have chosen the Mir Space Station as a first step toward their goal, and plan to keep Mir "alive" by employing a tether.

Satellite manufacturers are frequently challenged with the trade-off between long and short satellite life spans. On this topic **B. Madauss** stated that manufacturers try to comply with the costumers' needs, which constantly evolve, by orienting their designs toward a life span of fifteen years.

*M. Rycroft (ed.), The Space Transportation Market: Evolution or Revolution?*, 43–44.
© 2000 *Kluwer Academic Publishers. Printed in the Netherlands.*

**J. Friis** briefly commented on the services offered by Final Analysis, Inc.. He explained that the company plans to launch a constellation of satellites to allow data transmission. A network of ground stations will be established to provide tracking of assets. The satellites will be launched six at the time and be compatible with various launch vehicles. In addition, the company will provide secondary payload launches.

The question of Ariane 5 competing with Ariane 4 was raised; **P. Rudoph (Arianespace)** commented from the floor that Arianespace does not sell rockets but rather offers a complete service to its costumers. He made the analogy of airlines, where the customers select the airline but do not put emphasis on the particular type of aircraft used.

**P. Smith** was questioned about the procedures which a company must follow to certify a vehicle designated for space tourism. She pointed out that governments are not ready to support such activity as there are no regulations established. For instance the FAA has not approved any safety standards regarding reusable launch vehicles.

Nowadays, the tendency of the market is toward mergers and joint ventures (e.g., Boeing and Lockheed-Martin in the USA). Based on **R. Khadem's** personal point-of-view), mergers are beneficial to the customers if they lead to improved reliability. **S. Glynn** pointed out that competition is an important factor since it helps to keep the price of the launch vehicles down while increasing their reliability. In the case of a failure, the costumer has the option of selecting another launch vehicle.

The involvement of governments in the development of new launch vehicles was also discussed. **P. Collins** mentioned that the government's participation is required to absorb the technical risks and ensure faster development; that was the case for the aviation industry at the beginning of the twentieth century. **R. Tumlinson** added that the DC-X is an excellent example of government and industry cooperation in the design and technological development of a cost-effective space transportation vehicle.

**K. Bahrami** finalized the discussion by reminding the audience that the current market is limited by government regulations. It was suggested that there should be a balance between the government and industry requirements. In addition, **R. Tumlinson** proposed partnerships between national space agencies and industry to create new markets.

# Session 2

# Meeting the Needs of Future Markets — Launch Service Providers' Perspectives

Session Chair:

**G. Laslandes,** CNES-HQ, France

# Lockheed Martin Space Operations: Teaming and Outsourcing to Meet the Needs of Future Markets

**J. Honeycutt,** Lockheed Martin Space Operations 2625 Bay Area Boulevard, Houston, Texas, 77058, USA

e-mail: jay.honeycutt@lmco.com

**Abstract**
Globalization of the world economy shapes the space market place in numerous ways and commercialization has radically changed the traditional space markets. As the market sets the price, in order to remain both competitive and profitable, industry must maximize its efficiencies and control its costs while maintaining the quality of its products and services. Increasingly, teaming and commercial outsourcing are driving program development and efficiencies within the industry.

Through Lockheed Martin Space Operations (LMSO), the Lockheed Martin Corporation (LMCO) has developed teaming and commercial outsourcing expertise to meet successfully the needs of its commercial and government customers, both today and tomorrow.

## 1. Introduction

The on going globalization of the world economy continues to mold and shape the space market place in numerous ways. Commercialization and the adoption of commercial practice have radically changed the traditional space markets. The commercial space market has grown and evolved rapidly from its government roots and is itself still evolving into a sub-sector of the international telecommunications market place. The Government market place itself is not immune to change and is fast undergoing a wave of 'privatization'. Yet these market sectors still hold much in common, and meeting the needs of these future markets will require the same skills, talents, and abilities to provide the best products at the best prices. With the myriad of skills required to meet the customers' needs, it is rare that one company can work alone or, if so, has the efficiency and expertise needed to do so. As the market sets the price, in order to remain both competitive and profitable industry must maximize its efficiencies and control its costs while maintaining the quality of its products and services. Increasingly, teaming and commercial outsourcing are driving program development and efficiencies within the industry. Through Lockheed Martin Space Operations (LMSO), the Lockheed Martin Corporation (LMCO) has developed teaming and commercial outsourcing expertise to meet all the needs of its commercial and government customers.

Change is the only constant. The evolution of our market place and our end customers is transforming the space industry. For the first time in our history,

*M. Rycroft (ed.), The Space Transportation Market: Evolution or Revolution?, 47–53.*
© 2000 *Kluwer Academic Publishers. Printed in the Netherlands.*

the commercial customer base has surpassed its government predecessor in spending and influence. It is now commercial pressures that are driving our market place and hence our industry, as even the US Government market sector is moving towards 'privatization'. Following the US aerospace mergers, 'vertical integration' is now also the name of the game in Europe. The function of our industry remains the same, yet the market is continually evolving. The technical innovation that drove the genesis of the space industry has become static. Our products, while keeping pace with technological improvements, remain basically the same. Our core function as an industry is still to design, build, launch and operate satellites and spacecraft. In essence, the technology and products which we are dealing with have changed little. A satellite is still a satellite. A launch vehicle is still a launch vehicle. However, it is in the way that we are doing business and in the way that we meet the needs of our market place that we are now seeing revolutionary change. Our technology is mature; now, our services must evolve. If we are to meet successfully the needs of future markets then we must recognize and embrace these changes. The progression from government to commercial business is proving to be just the beginning, just the catalyst, for the accelerating changes around us. Lockheed Martin (LMCO), especially Lockheed Martin Space Operations (LMSO), is uniquely positioned to gain a rare perspective and grasp on changes in the market place, and in a position to act accordingly.

Lockheed Martin Space Operations (LMSO), based in Houston, Texas is the world's largest spacecraft and satellite operator, combining the joint space operations for both government and commercial sectors. Through our Consolidated Space Operations Contract (CSOC), we operate all of NASA's satellites and spacecraft, from low earth orbit (LEO) to deep space. Commercially, LMSO supports the entire constellation of Globalstar satellites. LMSO actively supports the US Space Shuttle program along with all US commercial launch activities. We are also working with the end users of the satellites in orbit, both in data services and remote sensing, our industry's ultimate customers. Through our various business activities Lockheed Martin has the opportunity to gain unique insight into the entire range activities in the space industry, through both sides of the commercial and government marketplace. Lockheed Martin supports all of the major market players in all of their functions. We have our fingers on the pulse of the market place.

## 2.　The Opportunities for Lockheed Martin

All involved in the aerospace industry are well aware of the corporate unions that have taken place in the US and are now taking place in Europe, as various market players merge and consolidate to form partnerships with former competitors. We have witnessed the mergers of Lockheed Martin and its heritage companies of Lockheed, Martin Marietta, Loral, and others. The same is also true for Boeing and is now true for the European Astrium. With the commercialization of space our industry's customer base has fundamentally changed. We are now witnessing the results of our attempts to adapt to the new realities of the marketplace and to meeting the demands and expectations of our customers. For example, the drive to mergers comes in part from market demand for 'turn key' 'one stop shop' services coupled with the drive for competitive pricing. It is market demand both from government and commercial customers, that has driven this move to purchase services at competitive prices.

At LMSO we see the international marketplace driving competitive behavior. If the needs of today's market are driving this behavior, what behavior will the future market drive? We must understand how this is happening today to have a better understanding of how to meet the needs of our future market. Arguably, the predominant factors in the customers' choice of either satellite or launch provider is based upon three main criteria—reliability, schedule, and price. Out of these, price is often the determining variable of success in a competitive business situation. Schedule and reliability are competitive issues, but for the most part they follow the industry standards. However, price is both determined and driven by the market place. In order to remain competitive we must meet the market price, but in order to remain profitable we must control our costs.

The US Government is rapidly moving to initiate commercialization and privatization policies in the space arena. NASA is seeking fixed price and price for service contracting in order to drive efficiencies. Acting more as a commercial customer, NASA is looking to get the best price for its purchases and to draw the best value from its assets in order to gain most from its investment. Opportunities arise daily, as new NASA assets and facilities become available for privatization. Contracting practices standard to commercial industry, but new to the US Government market, are being implemented for the first time. Here industry is leading the way forward with change, and Lockheed Martin is at the forefront of these initiatives.

In addition, the space market place has evolved in itself and this is driving new behavior and practices. The market has grown from government-based national markets, of limited value, almost overnight to a global competitive commercial market place. In order to win business in this new market place we have to be competitive; to be competitive we must be able to meet the market price. In order to survive we must be able to make a profit. To make a profit we must be able to control our costs. Meeting the needs of the market and hence our customers' needs are fundamental to business success.

One way to control our costs is to look for economies of scale and / or partnerships. The push for such economies has been a driving cause of the corporate consolidations which we have seen in the space industry recently. The launch industry and the ability to offer 'turn key' services are an excellent example of such consolidations driven by the needs of the market place. Governments procured space products and services from many different sources, a time consuming and costly process. It soon became apparent that the commercial customer preferred to make one bulk purchase of all necessary products and services. Whereas a government would run a series of competitive procurements, the commercial customer would seek one provider through which all services could be bought in a 'turn key' manner, i.e., the design, manufacture, launch, and operation of their product would be integrated by one supplier. In effect this is a 'one stop shop' approach to purchasing.

This proved to be a challenge to the existing market players in the launch and satellite services market that had been used to focused selling of their specific products. How would they integrate their sales activities? Initially, informal relationships that existed in the market place were solidified and formed into alliances with 'bulk buy' arrangements being made between launch providers and satellite manufacturers. An example is the well publicized arrangements made between Boeing and Hughes in order to enable such 'turn key' sales opportunities. Lockheed Martin was the first company in the world to be able to offer such services totally 'in house' through its International Launch Services (ILS) joint venture, its Athena Launch Vehicle program, its commercial satellite manufacturing and its space operations divisions.

Yet it soon became apparent that greater customer service, coupled with cost savings and economies of scale, and hence greater profits, could be found through consolidation, merger, and partnership. Following Lockheed Martin's lead, Boeing and Astrium have consolidated, purchased, and merged various companies and divisions in order to improve their market position. Thus Boeing and Astrium are now in a position also to offer 'one stop shop' 'turn key' services to the market place. The needs of today's marketplace are driving industry's actions and behavior.

However, Lockheed Martin has recognized, and very much believes, that far more than mergers and consolidations are needed to meet successfully the needs of today's and tomorrow's marketplace. Even considering Lockheed Martin's extensive corporate capabilities, products, and service offerings, often more is needed to satisfy the needs of the customer and the market. The global market has grown so fast and has such varied requirements that rarely can one company meet all of its needs. Lockheed Martin chooses to partner and build teams within the industry to enhance both its own and its team mates' abilities to win business and to succeed in the market. Lockheed Martin has made partnering and team-building a core approach to its successful business.

## 3.    The Current Partnerships of Lockheed Martin

In the launch industry, Lockheed Martin is successfully teamed with Boeing to provide NASA with cost-effective human spaceflight through the United Space Alliance (USA) as the contractor for NASA's Space Flight Operations Contract (SFOC). The customer, NASA, needed a more cost-effective approach to its launch activities and initiated the SFOC. In order to meet the customers' needs, Lockheed Martin partnered with Boeing (Rockwell), a decision driven both by the market and the customer.

In the international expendable launch vehicle market, Lockheed Martin has also led the way in teaming and partnering to meet market demands and needs. International Launch Services (ILS) was formed in June 1995 with the merger of Lockheed Martin Commercial Launch Services, the marketing and mission management arm for the Atlas vehicle, and Lockheed Khrunichev Energia, International, the marketing and mission management arm for commercial Proton launches. To date, ILS has proven to be a successful and profitable example of partnering and teaming to meet market needs. Lockheed Martin Intersputik is also a prime example of Lockheed Martin's success in international partnering.

At LMSO, we successfully partnered with Honeywell (formerly Allied Signal Technical Services) in building the team that won and now operates NASA's Consolidated Space Operations Contract (CSOC), a $3.4 billion contract managing all of NASA's space operations and communications needs. Through the CSOC contract we operate and manage facilities at ten locations around the globe working with every NASA center; we manage the operations of all NASA's satellites. Also, through CSOC, we support the commercial launch industry directly in range services utilizing NASA's TDRSS satellites. Our success in winning this business was wholly dependent upon our ability to build the right team to meet the needs of the market.

International partnerships are also a focus for LMSO where we have built relationships within the international business community. From our work around the world commercially supporting the Globalstar constellation of 48 communications satellites, to our unique partnership with Kongsberg of Norway, we have always sought to exceed in team building and finding the right partners for success. This trend will only continue in the future as our markets grow and evolve. For future success, others may wish to establish partnerships and team up with Lockheed Martin.

## 4.    The Future

As the marketplace drives our needs, as our very corporate structures change and adapt to the needs of the market, it is valid to ask what our future market will look like. Will our customer base stay the same? How might it change? What effect will changes have upon us? How must we adapt to these changes to be successful? The last decade has seen incredible changes sweep through the space industry as our customer base moved from government to commercial. The way in which we do business has fundamentally altered in response to this change. This change is vital and on going. The commercialization of our industry is by no means over and some would argue that it is only just beginning, that what has happened to date is merely the catalyst of future change. From our unique position serving both government and commercial customers at all levels of the industry, we at LMSO are able to discern some, but by no means all, of the coming changes.

Lockheed Martin is taking a holistic approach. The launch industry, the satellite industry, and the space operations industry have merged in response to market demands for 'turn key' 'one stop shop' services. Customers now can come to Lockheed Martin, or go to our competitors, for any or all of their launch, satellite, or operational needs. It is rare that there are separate markets now for such products, especially for geostationary orbit. Our products are staying the same, but our customers are changing.

It could also be said that the international telecommunications industry has now become the prime driver of needs and requirements for the commercial market place. The commercial space industry has become a hardware provider for the telecommunications industry, a sub-set of this larger market place. Our customers have evolved from government agencies to commercial satellite operators to international telecommunications service providers. The character, and hence the needs, of our customers is fundamentally changing. This process will only accelerate. In the future, will we be selling 'turn key' services to Microsoft or AOL alongside France Telecom, AT&T, and SkyTV?

E-Commerce Business to Business (B2B) activity is about to revolutionize our marketing, sales, and procurement activities. B2B solutions will radically change our cost centers and will thus effect competitiveness. Will our customers bid online for their 'turn key' services? What will our customers expect? Lockheed Martin has not only recognized this growing trend, but has successfully acted upon it. In January 1999, through our CSOC program in Houston, Lockheed Martin successfully initiated and executed NASA's, and the US Federal Government's, first B2B Internet-based procurement. LMSO personnel secured in orbit data services for the Triana mission using an Internet procurement process. A process that would traditionally have taken from three to six months and cost over US $ 500,000 was conducted in less than three weeks at a cost of less than US $ 30,000. Lockheed Martin is also on the cutting edge of international e-commerce 'B2B' activity. Through Commerce One and in concert with Boeing, Raytheon, and British Aerospace Systems, Lockheed Martin is initiating the world's first Internet trading exchange for the global aerospace industry. This groundbreaking use of the Internet will drive efficiencies and change within the market, and will define the future market itself.

One thing is clear, in order to succeed in meeting the needs of our future markets, industry must identify and embrace change, from e-commerce to privatization. Success in future markets will come to those that adopt such policies and practices. Lockheed Martin is embracing the future. Lockheed Martin understands the need for competition, while appreciating the strengths of team building and partnerships. Lockheed Martin is always looking for partners and new ventures. Lockheed Martin has the experience and the vision to meet the needs of the space markets of today and tomorrow.

# Long March Launch Vehicles — Responsive to both Domestic and International Market

**R. Gao,** Space Division, China Great Wall Industry Corporation, 30 Haidian Nanlu, 100080, Beijing, People's Republic of China

e-mail: rfgao@cgwic.com

**Abstract**
China developed its own Long March series of launch vehicles; it launched the very first China-built satellite, Dong Fang Hong 1 (DFH-1), on April 24, 1970, and so becoming the fifth country in the world to launch its own satellite using its own launch vehicle. When China put the Long March launch vehicles onto the international market on October 25, 1985, there had been 12 Long March launches and the launch vehicles made available were Long March 3 and Long March 2C. It was the Long March 3 launcher that was first used to launch the U.S.-built commercial communications satellite, AsiaSat-1. In response to market demand and after an analysis of the whole market situation, China started to develop new launch vehicles. The Long March 2E launch vehicle was developed when the major satellite manufacturers were proceeding with the new generation of larger telecommunications satellites, and there were upper stages remaining to be launched after the phase-out of the Space Shuttle for commercial launches in August 1986. As the test flight of the Long March 2E succeeded and the development of a larger launch vehicle for GTO missions proceeded as expected, the most powerful launcher in the Long March family, Long March 3B, was proposed to the international market. At the time of its development, the standard GTO capability of Long March 3B was **4,800** kg. After its first successful flight, its GTO performance was raised to 5,000 kg and now **5,100** kg is offered to customers. The Long March 3B outperformed the available U.S. launch vehicles for commercial launchers — Atlas and Delta. Unfortunately, the commercial launch services business has now been circumvented with quotas, sanctions, issues related to export licenses for commercial communications satellites, changes of jurisdiction over commercial communications satellites, etc.. For the future of the Long March launch vehicles, it is clear — both from the market, and technical aspects — how to respond to the demand.

## 1.    History

April 24 this year is the thirtieth anniversary of the launch of first China-made satellite, on the Long March 1 (LM-1) launch vehicle. This launch made China the fifth country in the world to launch its own satellite, using its own launch vehicle. China started to establish its space industry in October 1956. On May 17, 1958, the then state leader Chairman Mao made the political statement and commitment that "we too must build artificial satellites." This commitment at the very top and the decision by the Chinese government to develop its own satellites and launch vehicles in the same year ensured support from various governmental and industrial levels for the development of both the first artificial satellite and the Long March launch vehicle. Like most other expendable launch vehicles in the market, the early Long March launch vehicles originated from missiles. After the first launch of LM-1 in 1970, various Long

55

*M. Rycroft (ed.), The Space Transportation Market: Evolution or Revolution?*, 55–61.
© 2000 *Kluwer Academic Publishers. Printed in the Netherlands.*

March launch vehicles for experimental, scientific, and applications satellites have been developed to meet the launch demand, both at home and abroad.

The first Long March launch vehicle for applications satellites was the two stage LM-2C using storable propellants for the recoverable payloads. After four consecutive successful launches of LM-2C, the three stage launch vehicle, LM-3, was developed. LM-3 basically used LM-2C as its first two stages and its cryogenic third  stage was newly developed. The maiden flight of LM-3 on January 29, 1984, failed. About 70 days later, LM-3 successfully sent the first experimental GEO communications satellite into orbit on April 8, 1984. Both LM-2C and LM-3 were developed to launch domestic Chinese satellites. With more and more progress made in China in all sectors after the "opening to the outside" policy was carried out in 1978, China declared — on October 25, 1985 — that it would put its Long March series of launch vehicles onto the international launch services market. And the China Great Wall Industry Corporation was delegated the task of doing the business development and marketing for the Long March launch vehicles. At the time the announcement was made, there had been 12 Long March launches, all for the domestic satellites. China positioned its Long March launch vehicles as "supplementary to the international launch services market". The first commercial payloads were two devices onboard a Chinese recoverable spacecraft and the Swedish scientific satellite, Freja.

The timing of the Long March rockets to enter the international launch services market was just right. From August 1985 to March 1987, failures of both the Space Shuttle Challenger and major western expendable launch vehicles (ELV's) led to the grounding of some launch vehicles for an extended duration and to the shortage of launch opportunities in the market (the Launch Crisis in the mid 1980's; see Table 1). When the Reagan administration decided on August 15, 1986 to use the Space Shuttles solely for governmental and military missions, all the major US expendable launch vehicle manufacturers were in the early stages of a recovery from the phase-out and from commercialization policies for US ELV's. At that time, launch vehicles from the former Soviet Union were still prohibited from competing freely in the international market and Ariane, although it was still under further development in both reliability and performance, benefited most from the US policy change. The commercial communications satellite development gained momentum soon after the Launch Crisis. In October 1987, Hughes announced its plan for new generation telecommunications satellites, HS601, during Geneva Telecom '87. Other major satellite manufacturers followed and promoted their larger platforms for the new generation of telecommunications satellites. In response to the market demand for the launches of the new generation satellites, China Great Wall introduced the LM-2E launch vehicle onto the market. The intent was to use LM-2E and a US upper stage remaining from the Space Shuttle missions to launch HS601 or any other larger satellite. The LM-2E launch vehicle enjoyed its sound heritage from flight proven hardware and technologies. It used LM-2C as the core stages and incorporated four strap-on boosters developed from the LM-1. The timely introduction of the LM-2E launch vehicle led to the award of a contract from Hughes for the launch of two HS601 satellites for the AUSSAT B (later renamed Optus B) satellite program in November 1988, when the LM-2E launch vehicle was still in the development stage. In July 1989, AT&T awarded a contract to GE Astro Space for three Telstar 4 satellites based on the new GE7000 platform. The selection of HS601 and GE7000 by renowned operators both represented and confirmed the trend in commercial communications satellite development — multiple payloads, higher power and larger capacity. This new development, in turn, demanded larger launch vehicles to put the larger satellites into orbit; and LM-2E became one of a very small number of launch vehicles in the market.

| Launcher | Launch Date | Satellite(s) |
|----------|-------------|--------------|
| Titan 34D | August 28, 1985 | Keyhole 11#7 |
| Ariane 3 | September 12, 1985 | Spacenet 3/ECS 3 |
| Challenger | January 28, 1986 | TDRS-B/Spartan-Halley |
| Titan 34D | April 18, 1986 | Keyhole 9 Big Bird #20 |
| Delta 3914 | May 03, 1986 | GOES-G |
| Ariane 2 | May 31, 1986 | Intelsat V F14 |
| Atlas-Centaur | March 26, 1987 | Fltsatcom 6 |

**Table 1.** Launch failures from August 1985 to March 1987

## 2.    Commercial Long March Launches

The successful launch of the Hughes-built AsiaSat-1 satellite on LM-3 represented the entry of the Long March into the international market. The successful launch of the first HS601 on LM-2E further consolidated the Long March's position in the international market. As more and more marketing efforts were made, it was found that the two stage LM-2E launch vehicle for low Earth orbit missions, originally intended to couple with the upper stages for Space Shuttle launches for GTO missions, was sometimes too troublesome for some customers. To simplify the technical coordination for the satellite launch program, EPKM, a solid fuel perigee kick motor compatible with LM-2E for a GTO mission, was designed and developed; contracts for the launch of AsiaSat-2 and EchoStar-1 using LM-2E/EPKM were awarded. The successful launches of AsiaSat-2 and EchoStar-1 in November and December 1995, respectively, further enhanced the overall Long March launch capability. The LM-2E launches, are listed in Table 2.

| No. | Launch Date | Payload | Satellite Maker | Customer |
|-----|-------------|---------|-----------------|----------|
| F1 | July 16, 1990 | DP/Badr-A | SUPARCO | SUPARCO |
| F2 | August 14, 1992 | Optus-B1 | Hughes | Optus |
| F3 | December 21, 1992 | Optus B2 | Hughes | Optus |
| F4 | August 28, 1994 | Optus B3 | Hughes | Optus |
| F5 | January 26, 1995 | Apstar-2 | Hughes | APT |
| F6 | November 28, 1995 | AsiaSat-2 | Lockheed Martin | AsiaSat |
| F7 | December 28, 1995 | EchoStar-1 | Lockheed Martin | EchoStar |

**Table 2.** LM-2E launches

In the late 1980's, China started to develop new generation communications satellites for its domestic market. As the available LM-3 was too small to launch these larger satellites, LM-3A, a three stage launch vehicle, with a cryogenic third stage, was developed. After the successful demonstration flight of the LM-2E, a new version (the heavy-duty launch vehicle — LM-3B) was proposed to customers in the international market. Explained simply, LM-3B could be taken to be LM-3A as its core stage with four strap-on boosters from the flight proven LM-2E or, in other words, LM-2E with the cryogenic third stage of LM-3A. The heritage of the LM-3B was clear from its configuration.

When LM-3B was proposed to customers, few ELV's of the same class were available in the market. Boeing had not yet decided to develop its new generation launch vehicle and General Dynamics was upgrading its Atlas II series, and the development of the Atlas IIAR (Renamed Atlas III) was still at a very early stage. The technical approach taken for the development of LM-3B was endorsed by Intelsat. A contract to launch an Intelsat VIIA satellite on LM-3B was awarded to China Great Wall in April 1992 when both LM-3A and LM-3B were at different stages of their development. Unfortunately, after seven LM-2E and two LM-3A flights, the maiden flight of LM-3B carrying the Intelsat 708 satellite failed on February 15, 1996. The failure analysis and corrective actions achieved good results later. Four consecutive LM-3B launches were successful and put satellites into either a super-synchronous transfer orbit or a GTO with lower inclination. The launch capability of LM-3B was also raised from the 4,800 kg at the beginning to 5,000 kg. Now 5,100 kg is offered to customers for a typical GTO mission. The LM-3A and LM-3B launches are listed in Table 3.

| Launcher | Launch Date | Payload(s) |
|----------|-------------|------------|
| LM-3A  F1 | February 8, 1994 | Domestic Satellites |
| LM-3A  F2 | November 30, 1994 | DFH-3 F1 |
| LM-3B  F1 | February 15, 1996 | Intelsat 708 |
| LM-3A  F3 | May 12, 1997 | DFH-3 F2 |
| LM-3B  F2 | August 20, 1997 | Mabuhay |
| LM-3B  F3 | October 17, 1997 | Apstar IIR |
| LM-3B  F4 | May 30, 1998 | ChinaStar-1 |
| LM-3B  F5 | July 18, 1998 | SinoSat-1 |
| LM-3A  F4 | January 26, 2000 | ChinaSat-22 |

**Table 3.** LM-3A and LM-3B launches

The development of the Long March series launch vehicles has been closely related to the market demand from both domestic and international markets. In the last thirty years since the first LM-1 was launched, there have been sixty launches involving several types of Long March launch vehicles. The average number of launches has increased from less than one launch a year in the first fifteen years to slightly more than three launches a year in the second fifteen years (see Fig. 1). Since its last failure in August 1996, eighteen consecutive successful launches have been achieved, for both domestic and commercial satellites. The premium rate for the launch insurance has been reduced substantially because of the most recent track record. And with the improvements of Long March services and facilities, a more and more positive market response has been expected; in reality, the market has not responded in such a way.

## 3.    Political Considerations

Since Long March came into the market, the international  launch services market has become highly  regulated  and  protected.  In 1988, when President Reagan decided to allow US-built commercial communications satellites to be launched on Long March launch vehicles, the Chinese government was requested to sign bilateral memoranda of agreement (MOA's) on satellite technology safeguards, liability for satellite launches and regarding international trade in commercial launch services. The MOA regarding international trade in commercial launch services stipulated the quota, pricing policy, etc.. Similar MOA's were signed between the US government and the Russian government or Ukrainian government. During the performance of these bilateral MOA's, suspensions and sanctions were imposed on the China launch services provider which prohibited the shipment of US-built satellites to China for launch.

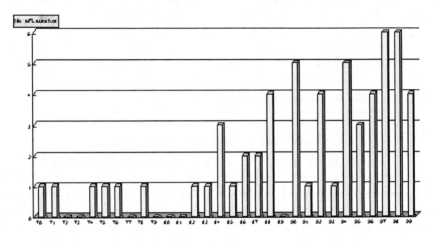

**Figure 1.** Long March launches in last three decades

After the release of the Cox Report in December 1998, the new ruling (becoming effective on March 15, 1999) changed the licensing authority from the US Department of Commerce to the Department of State, and more time consuming and restrictive licensing for all commercial communications satellites had been put in place. All these actions have, to certain extent, had a negative impact on the business development and marketing efforts for Long March launch vehicles. Even though the export of US commercial satellites and satellite components has decreased since the tight export control was carried out, the status of US expendable launch vehicles might be the cause of all the concerns. Since the withdrawal of the Titan III from the international launch services market, the Delta and Atlas series have evolved at a relatively slow pace. The veteran Delta II has enjoyed a very good track record, but it is too small for the larger commercial communications satellites and has been used for a specific market segment. Launch performance of the newly developed Delta III is still less than 4,000 kg, and it still needs time to prove itself. The successful maiden flight of the well-designed Atlas III on May 24, 2000 has just increased the Atlas GTO performance to 4,500 kg, still below the launch capability of other existing and competing launch vehicles, including LM-3B.

Export controls have not substantially hampered the development of commercial communications satellites. Since the introduction of the HS702 during Geneva Telecom '95, the first HS702 has been successfully launched and several more HS702 satellites ordered. Other satellite manufacturers have also started to market their new larger platforms in order to compete. The launch services providers have either taken the lead in developing larger launches or responded with a plan to build a more powerful launch vehicle. After four consecutive successful launches of LM-3B, a larger version of the Long March launch vehicle based on LM-3B for HS702 class satellites is under study. The areas of improvement are the payload fairing and the launch performance for the larger commercial communications satellites.

## 4.   Conclusion

In conclusion, the Long March launch vehicle developments have been driven by the needs from both domestic and international markets, and new developments are still going on. The approaches which have been taken are evolutionary.

Looking at today's market situation, China Great Wall still positions its launch services with the Long March launch vehicles as "supplementary to the international launch services market." The commercial satellite launch market is still highly regulated and protected. For the future development of the Long March launch vehicles in the international launch services market, many obstacles, hurdles and roadblocks need to be removed.

# The Future is Not Like the Past, or How the Cost of Access to Space may Change the Market

B. Parkinson, Astrium Ltd, Gunnels Wood Road, Stevenage SG1 2 AS, UK

e-mail: bob.parkinson@astrium-space.com

### Abstract

Launch vehicles have long development times and even longer operational lives. Prediction of future payload requirements is therefore difficult. The prediction of future payload requirements for "next generation" launchers has tended to be an extrapolation of current trends. Performance of launch vehicles has tended to be calculated on a cost per kilogram in orbit. As a consequence (expendable) launchers have tended to grow larger, with an accompanying assumption that satellite masses will also grow. The geosynchronous satellite market, currently the primary commercial launcher customer, has followed this pattern. However, some disruptive trends may be detected in which the "rules of the game" change. While "constellation" communications systems have yet to prove successful, these demanded large numbers of relatively small satellites in multiple orbits, with the capability of replacing individual losses at low cost. At still smaller sizes, the capabilities of mini- and micro-spacecraft have been growing, but the development of this market is inhibited by the cost of launch. Currently this can be two to three times the cost of the spacecraft. Further new applications may exist if both cost and accessibility can be improved. Opportunities to meet these (largely hidden) markets may come through providing secondary launch opportunities on large, cost-efficient current launchers, or possibly through the development of small re-usable vehicles.

## 1.    Introduction

Launch vehicles are designed to the size of the spacecraft they expect to carry. At the same time, since it costs as much to launch a half-full vehicle as a completely full one, satellites tend to be designed to fit the launch vehicles they expect to use. There is thus a paradoxical relationship between launchers and their payloads. If a suitable launch vehicle is not available, or the price is wrong, spacecraft will simply not be proposed. Under such circumstances new applications will not be thought about, and new markets will go undiscovered.

Faced with a period of five to ten years to bring a new launch vehicle into operation, launcher developers have tended to rely on extrapolations of past trends to predict the future. Fig. 1 shows the trend for "average" GTO satellite masses, derived from Arianespace data, extrapolated into the future. It is on the basis of such extrapolations, carried out in depth, that decisions were made to move from Ariane 4 to the larger Ariane 5. But one of the basic rules of futurology, carried any distance into the future, is that "the future is not a simple extrapolation of the past."

*M. Rycroft (ed.), The Space Transportation Market: Evolution or Revolution?, 63–71.*
© *2000 Kluwer Academic Publishers. Printed in the Netherlands.*

In addition, there has been a steady tendency for established launch vehicles to grow in size. In part this is because launch vehicle designers tend to measure performance in terms of cost per kilogram delivered into orbit. Since the additional costs of a "stretched" vehicle are often small in comparison with the payload gain, larger (expendable) launch vehicles tend to have lower cost per kilogram than smaller vehicles. Fig. 2 shows the growth in launch capability into LEO for the Delta family since its inception. From a first launch orbiting just 56 kg in 1960, the Delta vehicles have evolved to more than 5 tonnes capability, and the Delta III/Delta IV developments aim to more than double this.

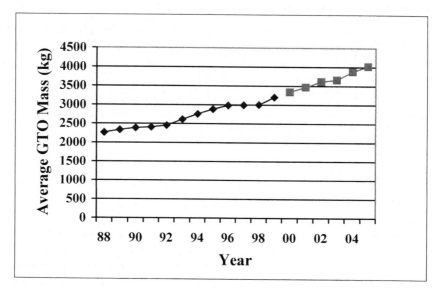

**Figure 1.** Growth in average GTO satellite mass with time

Some adaptability can be built into these large vehicles by adopting flexible build concepts with various "strap-on" configurations (Ariane 4, Delta III/IV, Atlas V) and/or using dual launch systems (Ariane) to "mix and match" smaller satellites. But the overall trend is to "bigger and better".

Christensen [Reference 1] has shown how this sort of behaviour, showing a steady drift to bigger and higher performing systems is typical of what he describes as "sustaining technological change", when the rules of the game are well established and the customers well known. But he also points out that this stable market situation can be disrupted. In very simplified terms this can occur when:

• The mainstream performance is more than adequate to meet the market requirements, and

• A new player starts to open up a related market which cannot be served by the primary suppliers, and which changes the rules of the game (the definition of "performance".

The object of this paper is to see whether there are markets which are not well served by current launchers, and how these might develop.

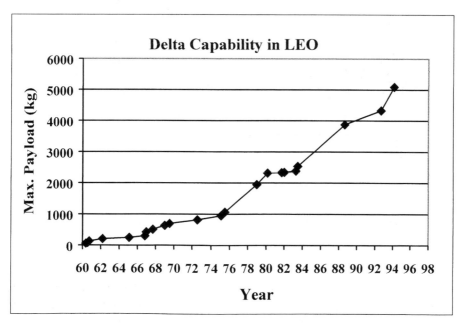

**Figure 2.** Growth in LEO payload capability of Delta launch vehicles

## 2.   Small Satellites

Despite attempts to develop small launchers, the small satellite market (<
800 kg) is not well served by current launch systems. Developments in small
satellite technology and management approaches have meant that it is possible
to produce satellites with attractive capabilities at low mass and cost. Microsatellites (< 120 kg) are of interest to universities and small institutions, and can
be built for ~ $ 2 million. Mini-satellites (200 – 500 kg) have practical
applications, and can be constructed at costs in the region of $ 5 million.
However, at payload masses of below about 2000 kg the specific cost of launch
rises rapidly (Fig. 3), and typical launcher costs are in excess of $ 14 million.
Since construction costs are often "in house" while the launcher is a "bought
out" item, this disparity inhibits market growth.

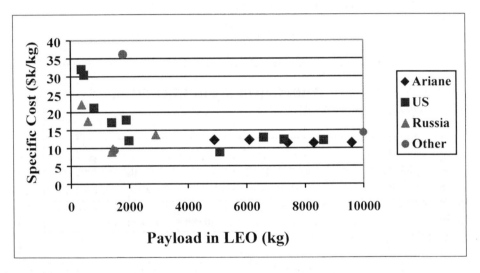

**Figure 3.** Specific launch costs (into LEO) for current launch vehicles

Micro-satellites, on the whole, have found it easier to get affordable rides
than mini-satellites. Micro-satellites can ride relatively easily as "piggy-back"
payloads on larger launchers. In addition since many come from academic
institutions, they find it relatively easy to get favourable treatment from
agencies responsible for the primary launch. As can be seen in Fig. 4, number of
micro-satellites (< 120 kg) launched over the past few years has grown strongly

[Reference 2]. However, as numbers increase, available low cost launch slots may become more difficult to find.

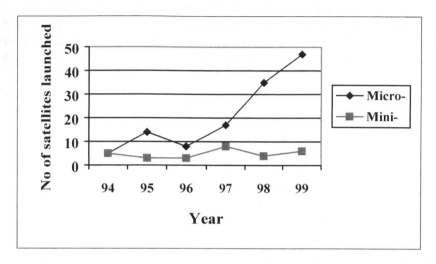

**Figure 4.** Growth in numbers of micro- and mini-satellites launched

Some underlying upward trend can be detected even with the mini-satellites. But here the market appears to be seriously inhibited by the lack of launch opportunities. With improvements in technology, mini-satellites are potentially attractive for low cost commercial applications. If sufficient "slots" could be found at the right price, the market could be expected to grow more rapidly than expected from past extrapolation.

Fig. 3 indicates that the economic solution to providing for this market is not to develop small launch vehicles, but providing secondary launch opportunities on larger vehicles (> 2000 kg in LEO) at an economic per-kilogram rate. The effectiveness of dual launch for larger satellites has been demonstrated by the Matra Marconi Space SPELDA system on Ariane 4. A micro-sat capability will be introduced onto Ariane 5 with the ASAP-5 platform. Secondary launch capability for small satellites has been introduced onto Delta II with the Astrium-built DPAF unit. Initially it ought to be possible to find secondary rides with larger payloads, but as demand grows it should be possible to package multiple (diverse) small satellites using a combination of dual launch and dispenser technology.

## 3.    Reusable Launch Vehicles

Reusable launch vehicle (RLV) proposals are generally made on the basis that their introduction will reduce the cost of access to space, and hence encourage new business. General economic principles indicate that, if the cost of a service is reduced, the utilization of the service will increase at a faster rate, with the net result that the total money flowing through the system actually increases. It is difficult to predict where this new business will come from in the case of space transportation, but if the rule holds it means that there will be an expansion which dilutes the traditional (and hence extrapolatable) business areas of communications and government sponsored activities.

By their nature, RLVs will only deliver into low Earth orbit. Transport to higher orbits will require upper stages. As long as these stages are expendable, they will form a significant part of the price of reaching higher energy orbits, and there will therefore be a trade between stage performance and stage cost. In the early days of the Space Shuttle it was assumed that these stages would be high-energy cryogenic vehicles like Centaur. In actuality, cheaper, lower performance solid propellant stages were adopted. For future RLVs there could be significant interest in low thrust solar-electric or solar-thermal stages which would give much higher payload fractions in HEO — or allow the use of smaller RLVs (see below). Some possible performances for co-planar LEO-to-GEO transfer are shown in Table 1.

| Transfer Stage | Mass Fraction In GEO | Transfer Time (days) |
|---|---|---|
| Bipropellant, 2-stage | 25% | 1 |
| Cryogenic, 1-stage | 32% | 1 |
| SPT solar-electric | 36% | 66 |
| Solar-thermal | 40% | 20 |
| Ion propulsion | 43% | 100 |

**Table 1.** Typical performance of LEO-to-GEO transfer stages

A major contributor to RLV costs is the cost of developing the system in the first place. Since finance for these costs must be found "up front" there is a significant interest in keeping them as low as possible. If we make the simplifying assumption that the objective of the system is to transport a certain total mass into orbit, then while direct operating costs could be lower with a larger vehicle, overall costs per kilogram tend to rise (see Fig. 5). There is therefore a tendency to develop a modestly sized vehicle and fly one payload at a time. The only RLV in advanced development at this time — the Kistler K-1 — takes this route [Reference 3].

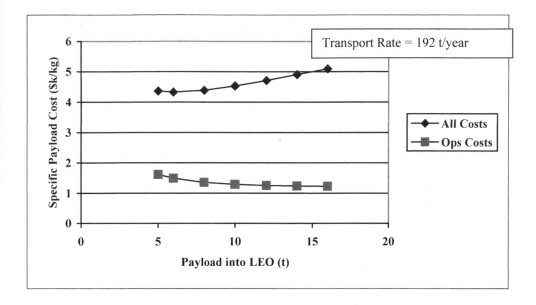

**Figure 5.** Typical costs per kilogram for an RLV transporting a given mass of payload into orbit each year (flight rate varies with vehicle size)

However, the amount of payload to be transported is not independent of vehicle size. The larger the vehicle, the greater the number of payloads it can accommodate. But a large vehicle has a higher unit launch cost and is therefore less attractive to smaller payloads. There is therefore a trade-off between available market and launch price per flight. With RLVs it is more difficult to accommodate a range of payload sizes by adapting the build with "strap on boosters" as is done with ELVs. As a consequence, RLVs may have to adopt not only new types of upper stage, but also new operational approaches such as on-orbit rendezvous of satellite and transfer stage, or to use the ISS as a transport node. New LEO markets may judge performance not only in cost-per-kilogram, but also in terms of readiness of access to orbit. Debris may also become an important issue, favouring recoverable systems generally.

Modestly sized RLVs are well suited to the LEO constellation market. The experience to date has not been encouraging for constellations, in part due to the high initial investment of setting up the system. However, developments in small-satellite technology coupled with a low cost of access to LEO could well start to change this pattern. Among other things, if it is cheap to replace a satellite, there need not be such emphasis on reliability and extended lifetimes. If we follow Christensen's model [Reference 1], the initial activity will involve specialist services not well served by heavy GEO satellites. But as the performance of the LEO satellites grows there may be a progressive take-over of the large GEO-satellite business.

Even modest RLVs will not solve the problems of the individual small satellite customer. The unit launch price will still be above $12 million (Kistler quote $ 17 million as an example). Such small satellites will still need to find rides as secondary passengers [Reference 4], or share the price of a ride by flying in clusters. If the market for mini-satellites grows, there is a possible opportunity here for a manoeuverable, high performance stage that could drop of a series of mini-satellites into a range of orbits from a single RLV launch.

## 4.    Conclusions

Current expendable launch vehicle developments have been towards larger vehicles delivering heavier payloads at an improved cost-per-kilogram in orbit. Some flexibility is being built in through the use of alternative vehicle builds and the use of strap on boosters, and through the use of dual launch systems. However, there appears to be a significant, although currently hidden market for micro- and mini-satellite launch services which is not currently adequately addressed by either the mainstream launchers or small launchers supposedly sized for this market. To achieve launch costs comparable with satellite costs will require providing secondary launch opportunities. As demand grows it may be possible to package multiple (diverse) small satellites using a combination of dual launch and dispenser technology.

Reusable launch vehicles may start with modest capabilities, possibly enabling LEO constellation projects. However, such vehicles still will not serve the small satellite market on a "one satellite, one launch" basis. It will still be necessary to provide secondary launch provisions on these vehicles. RLVs can be expected to place a new emphasis on flexible upper stage concepts, and some of these might be adapted to delivering clusters of small satellites into diverse orbits. Small satellite technology combined with modestly sized RLVs could ultimately displace large ELVs serving the heavy GEO market.

**References**

1.   Christensen, C. M.: *"The Innovator's Dilemma – When New Technologies Cause Great Firms to Fail"* Harvard Business School Press, Boston Mass., 1997
2.   Surrey Satellite Technology Ltd., Small Satellites Home Page: <http://www.ee.surrey.ac.uk/SSC/SSHP>.
3.   Mueller, G.E., Cuzzupoli, J., Kohrs, R. and Lepore, D.F.: *Building the K-1 Reusable Aerospace Vehicle as a Commercial Venture*, IAA-99-IAA.1.2.02, 50th International Astronautical Congress, Amsterdam, the Netherlands, October 4-8, 1999
4.   Lepore, D.F., Lai G. & Taylor T.: *Opportunities for Small Satellites and Space Research Using the K-1 Vehicle*, SSC'99-X-5, 13th Annual AIAA/USU Conference on Small Satellites, Logan, Utah, August 1999

# Starsem: A Euro-Russian Venture
# for Space Transportation

**P. Bonguet,** Starsem, Tour Maine-Montparnasse, BP n°30, 75755 Paris Cedex 15, France

e-mail: patrick.bonguet@starsem.com

### Abstract

The 1990's have seen tremendous changes in the satellite and space transportation businesses. Dealing with ever more demanding, less and less "space specific" customers and facing an increasingly aggressive competition, the launch services industry has to adapt very quickly to this evolving situation. This explains the implementation of global strategies resting on international alliances and technical improvements aiming at closely adapting the offer to the market trends. This context led world-leading companies and public entities to create, in 1996, the European-Russian joint venture Starsem, devoted to exclusive worldwide marketing and selling of Soyuz launch services. Meeting the specific needs of private customers and public agencies all around the world, the Soyuz launch system is very well adapted to launch missions of the "smaller, better, cheaper" era, including planetary missions. The performance and affordability of the launcher allow Starsem to offer an optimized solution to creating space projects of all kinds. Its reliability becomes a crucial asset when risks have to be minimized in order to cope with ambitious business plans or with scarce public resources.

The medium-class Soyuz launch vehicle is one of the most frequently used launch vehicles ever built, providing safe and reliable service for satellite payloads as well as for manned missions. To date, more than 1600 launches have been carried out. The core of the Soyuz family is a basic three-stage launch vehicle. The use of additional upper stages (Ikar, available today or Fregat, available later in 2000) enables the performance of the launcher to be adapted for the mission targeted and to complement perfectly the range of services provided by the Ariane family of launchers. A new version, Soyuz/ST, is under development and should be available from 2001. The launch site in Baikonur (Kazakhstan) features brand new state-of-the-art Starsem Payload Processing Facilities (SPPF) where the satellites are tested, fueled and mounted on the launcher in purpose-built clean rooms, as well as an international class hotel (Sputnik) to accommodate Starsem and customers' personnel. Leaders in space activities such as Space Systems Loral or the European Space Agency have already chosen Starsem. With six successful launches performed in 1999 for the Globalstar constellation, a promising year 2000 with two successful launches, a solid equity base and a solid commercial backlog, Starsem, symbol of Euro-Russian cooperation in space, has now become an essential player in the international launch services market.

## 1.  Introduction

In the early 1990s, changes in the geopolitical situation worldwide led the Russian authorities to seek ways to market a number of space systems developed by the Soviet Union. In this context, Aerospatiale was quick to establish links with the leading players in Russian space activities, including in particular RKA (the Russian Space Agency), and TsSKB-Progress (the Samara space centre), responsible for the development, manufacture and operation of the Soyuz launcher. It soon became clear that the Soyuz launcher could represent an extremely appropriate addition to the range of Ariane launchers

*M. Rycroft (ed.), The Space Transportation Market: Evolution or Revolution?, 73–79.*

manufactured by Aerospatiale Matra, and marketed and operated by Arianespace. This led to the formation of the Starsem company, the capital of which is held by Aerospatiale Matra (35%), Arianespace (15%), RKA (25%) and TsSKB-Progress (25%), for the purpose of marketing and operating the Soyuz launcher and various space systems.

## 2.    Starsem Services

From the time of the formation of the company, in 1996, the Starsem shareholders have aimed to make the company a focal point for European-Russian partnership in the space field. Apart from exclusive marketing of the Soyuz launchers in the world market, Starsem activities extend to all stages of the production process and operation of the launcher.

The company monitors the manufacture of launchers at the Samara plant, and participates in the development of dedicated versions designed to cater for the specific needs of customers (Ikar and Fregat upper stages, and the future Soyuz/ST launcher).

Following a substantial investment at the Baikonur cosmodrome, Starsem now has the necessary facilities for the preparation of customer satellites, and residential accommodation for customer personnel. Three western-standard clean rooms, for satellite preparation, fueling and integration, are now fully operational and were used throughout 1999 for the Globalstar satellites. The international class Sputnik Hotel, built by Starsem, was opened at the beginning of the year, and caters for the requirements of the technical teams working on the cosmodrome.

Apart from these basic technical aspects, Starsem activities have also required the creation of an appropriate regulatory framework, with particular reference to customs-related aspects. This objective has now been achieved, first with the implementation of the French-Russian inter-governmental space cooperation agreement and application of the rider to this agreement relating to customs duty and tax exemption, and secondly with the signature of an agreement concerning the protection of technologies associated with the launch of satellites incorporating sensitive components from the Baikonur cosmodrome, by the United States, Russian and Kazakhstan governments.

To provide its customers with a comprehensive service, Starsem has designed and introduced dedicated mechanisms designed to meet customer requirements to the fullest possible degree in the area of financing for the launch services provided.

Starsem illustrated its ability to provide a turn-key launch service during the Globalstar campaigns in 1999. Six launches were performed on behalf of Space Systems Loral sending into orbit half of the satellite constellation: ST 01 on 9 February, ST 02 on 15 March, ST 03 on 15 April, ST 04 on 22 September, ST 05 on 18 October and ST 06 on 22 November. The year 2000 began with two qualification launches with the Soyuz-Fregat launch vehicle configuration: ST 07 on 9 February and ST 08 on 20 March.

## 3.    Company Organisation

The company is organised around four core units. These are:

- A sales unit responsible for marketing, sales and monitoring customer contracts

- A launcher unit, the activities of which cover the monitoring of launcher production at Samara, the adaptation of launchers for specific customer missions, and technical management of new developments

- An operations unit, responsible for preparing and conducting launch campaigns, and adapting and managing the infrastructures required for launch operations

- A finance unit, dealing with the financial, legal and human resources aspects of company activities.

Starsem has a staff of about fifty, with a team of five based permanently in Baikonur, in compliance with the parity rules desired by the founders of the company. Starsem also maintains permanent links with its French and Russian partners involved in the implementation of its activities.

## 4.    Range of Vehicles

The Soyuz family is based on a three-stage launcher, a descendant of the vehicles used in the very early days of Soviet astronautical activities in the late 1950s. More than half of the total number of launches in the world have been performed with a Soyuz launcher. The addition of a reignitable fourth stage (also referred to as the upper stage) gives the launcher a degree of operational flexibility enabling it to undertake complex missions (change of orbital plane, change of altitude and orbit circularisation), as are now frequently encountered, particularly since the appearance of LEO satellite constellations.

## 4.1   Soyuz-Ikar

The simplest upper stage version, Ikar, is derived from the Kometa platforms used on more than twenty flights for Russian satellite programmes. The Ikar stage can be reignited up to fifty times. This makes it possible to place payloads of 4.1 tonnes (altitude 450 km, inclination 51.8°) to 3.3 tonnes (1,400 km, 51.8°) in a high altitude circular orbit, inaccessible to the three-stage version of the Soyuz launcher. The Soyuz-Ikar launcher has already flown six times for the ST 01 to ST 06 missions, each of which placed four Globalstar satellites into orbit. In-flight performance has proved to be fully in line with the predictions.

## 4.2   Soyuz-Fregat

The Fregat upper stage, operational since March 2000, is more flexible and more powerful. In order to maintain the maximum reliability of the Soyuz launcher, the Fregat upper stage incorporates systems and sub-systems which have already flown on Russian satellites and launchers. The propulsion system and the main engine have been used successfully for no fewer than 27 interplanetary probes. The Soyuz-Fregat launcher is able to place payloads of 5.0 tonnes (450 km, 51.8°) to 4.2 tonnes (1,400 km, 51.8°) into circular orbit. This vehicle is also capable of achieving Sun-synchronous polar orbits (2.7 tonnes, 800 km) and geostationary transfer orbits GTO (1.3 tonnes Cape Canaveral equivalent).

## 4.3   Soyuz/ST

The family of Soyuz launchers marketed by Starsem should be further extended in the year 2001, with the arrival of a new version of the launcher developed specifically for Starsem and designated Soyuz/ST. This programme will enable Starsem to acquire, very rapidly, a more powerful version of the launcher with increased fairing capacity, essential for retaining the company's market share, and penetrating the satellite system market as foreseen for a five-year horizon. In order to optimise development costs for larger fairings, Soyuz/ST will feature an Ariane 4 type fairing, a standard now adopted by all satellite designers for payload definition (see Fig. 1).

With the Fregat upper stage, the Soyuz/ST launcher will be able to place payloads of 5.5 tonnes (450 km, 51.8°) to 4.6 tonnes (1,400 km, 51.8°) into circular orbit and about 4.5 tonnes to Sun-synchronous orbit at 800 km.

**Figure 1.** Soyuz/ST

## 5.    Order Book

The telecommunications operator Globalstar (three times) and the European Space Agency (twice) have already placed orders with Starsem. The company has also established close links with numerous potential customers, including both government agencies and private operators, and can reasonably expect to sign new launch contracts in the short term.

The Globalstar launch contracts demonstrate the extent to which Starsem forms a complement to Arianespace. The specific nature of Starsem places the company in a particularly strong position to meet the substantial demand from LEO or MEO telecommunications satellite constellation operators. Starsem used the Soyuz-Ikar launcher for Globalstar flights, and this version will be flown again this year and in 2001, in each case carrying four satellites.

Starsem has also obtained two launch contracts from the European Space Agency. The four Cluster II satellites will be launched in pairs, by two Soyuz-Fregat launchers in mid-2000 and the 1.1 tonne Mars Express interplanetary probe will be sent to Mars atop a Soyuz-Fregat launcher from Baikonur in June 2003. For SkyBridge/Alcatel, 32 satellites will be launched for the SkyBridge constellation.

In parallel, the company has been engaged in intensive prospecting activities for marketing the Soyuz launcher and other space systems. The 1999 and 2000 launches, combined with the plans to develop Soyuz/ST, have strengthened Starsem's credibility. The company now has a solid commercial base, which should lead to substantial new business in the near future.

## 6.    Conclusion

The European-Russian partnership in the launchers domain, as represented by Starsem, now possesses a decisive advantage in world competition. Using a launcher of proven reliability and which is complementary to the Ariane range, with the backing of a strong production infrastructure and unequalled marketing expertise, and capable of offering an extremely advantageous price policy, the company has also enjoyed the favourable attention of the French and Russian governmental authorities since its formation. To draw the full benefit from this situation, and confirm its promising prospects, Starsem must now ensure its future development by action in two directions.

- The company must continue to implement the ambitious launch programme which is the consequence of its early successes ; the Starsem order book enables the company to schedule launches up to 2003

- The Soyuz launcher must evolve in technical terms, in order to meet customer demand in the fullest possible way, and that is the reason behind the development of Soyuz/ST. Furthermore, utilisation of an equatorial launch base could open up new opportunities by broadening the company's commercial product range.

Having demonstrated the reliability of the development work conducted by its Russian partners, and its capacity to conduct launch campaigns (see Fig. 2) at a sustained rate, Starsem must now prepare product enhancements to achieve optimum adaptation to market demand and perpetuate the advantages which it already possesses, in order to become an essential player in the launch services market.

**Figure 2.** Starsem flight 08, Soyuz-Fregat, March 20, 2000, 11:28 p.m., Baikonur local time

# The Rockot Launch System

**M. Kinnersley, P. Freeborn, K. Schefold,** EUROCKOT Launch Services GmbH, Flughafenallee 26, D-28199 Bremen, Germany

e-mail: mark.kinnersley@astrium-space.com

**Abstract**

Rockot can launch satellites weighing up to 1850 kg into polar, Sun synchronous (SSO) or other low Earth orbits (LEO). The Rockot launch vehicle is based on the former Russian SS-19 strategic missile. The first and second stages are inherited from the SS-19, the third stage Breeze has been newly developed and is characterized by a multiple ignition capability. The Rockot launcher is flight proven and marketed by Eurockot. Eurockot Launch Services GmbH was founded by DaimlerChrysler Aerospace of Germany (now Astrium) and Khrunichev State Research and Production Space Center of Russia (KSRC) to offer world-wide cost-effective launch services on the Rockot launch system. The extremely high flexibility of both the payload accommodation and the launch system allows a fast service - a nominal launch campaign lasts about 20 working days. The high degree of maturity of the launch vehicle and the launch system contribute to its competitively fair price.

Customer-oriented developments of the launcher include an enlarged payload fairing providing more space for satellites and a flattened equipment bay for accommodating several spacecraft. The Commercial Demonstration Flight (CDF) marks the debut of Eurockot Launch Service's commercial operations with this newly configured Rockot launch vehicle. The mission successfully launched two identical satellite simulators, SIMSAT-1 and -2, into a near polar orbit using a typical LEO mission profile from the newly commissioned Eurockot launch facilities at the Plesetsk cosmodrome in northern Russia on 16th May 2000.

## 1. Launch Vehicle Heritage

The Rockot launch vehicle uses a decommissioned RS-18 ICBM (NATO designation: SS-19 Stiletto) as its first two stages. The RS-18 was developed between 1964 and 1975 by the Chelomei Design Bureau, which later evolved into the Khrunichev State research and Production Space Center. A restartable, upper stage, called Breeze-K, had been developed by Khrunichev to make the launcher suitable for orbital launches. Three successful silo test launches have been conducted from the Baikonur Cosmodrome between 1990 and 1994, of which two were suborbital and one was orbital. The Breeze-KM stage is a structurally modified version of the original Breeze-K stage. It provides more payload space.

The commercial demonstration flight with the newest Rockot configuration was carried through from Plesetsk in May 2000. The launch was conducted out of a transport- and launch- container serviced during Launch preparation and countdown by a launch tower (see Fig. 1). Further details are given in Sections 4 and 5.

*M. Rycroft (ed.), The Space Transportation Market: Evolution or Revolution?*, 81–92.

**Figure 1.** Rockot launch service tower in Plesetsk

## 2.    Commercialization of the Rockot Launcher

To commercialize the Rockot booster and reach potential western customers, Khrunichev (KSRC) teamed up with Daimler-Benz Aerospace (DASA), now Astrium, in 1995 to form a joint venture named EUROCKOT Launch Services. KSRC, which provides the launch vehicle, refurbished the launch site and is responsible for conducting launch operations in conjunction with the Russian Military Space Forces. DASA contributed capital to complement the Rockot launcher launch site infrastructure in Plesetsk, to upgrade and refurbish the launch facilities. EUROCKOT is the single point of contact for the customer.

Marketing efforts began in 1997, and EUROCKOT has since signed contracts for several launches beginning in 2001. The most common potential payloads are satellites to make up LEO communications constellations or international small scientific satellites.

## 3.    Rockot Vehicle Design

### 3.1    Lower Stages

The first and second stages of Rockot are decommissioned RS-18 ICBM stages with about 80 tonnes of  propellant ($N_2O_4$ and UDMH) in the first stage and about 14 tonnes in the second stage. Four first stage closed cycle turbopump RD-0233 engines with, a single axis nozzle gimbal on each engine, provide the main thrust and steering capability. The second stage contains a closed cycle, turbopump engine designated RD-0235, and one Vernier engine with swivelable nozzles for three axis control. These stages have been flight proven more than 146 times and described in many articles.

## 3.2  Breeze Upper Stage and Payload Fairing

The new Breeze-KM is a structurally modified version of the original Breeze-K stage; they are functionally identical; the weight of the KM-version differs slightly from that of the K-configuration. The stage includes an oxidizer ($N_2O_4$) tank that encircles the main engine and shares a common bulkhead with the fuel (UDMH) tank.

The original conical equipment section has been flattened and widened to provide more room for the payload. The KM upper stage is no longer attached to the launcher at its base but hung within the extended transition compartment. Consequently, the fairing is now attached directly to the equipment compartment (see Fig.2).

The Breeze houses a multiple restartable, turbopump fed 20 kN main engine, a bipropellant pressure fed engine with four 400 N thrusters for ullage control and orbital maneuvers, and an attitude control system providing 12 x 16 N thrusters. The avionics are located in the equipment bay on top of the Breeze stage. The guidance, navigation and control system controls the vehicle during all stages of flight. It includes an inertial guidance system with a three-axis gyro platform, and a flight computer. Control commands are computed in three separate channels, with majority voting to correct errors.

The telemetry system includes onboard tape recorders to store in-flight data until they can be transmitted to an available tracking station. Vehicle tracking is facilitated by a beacon. The usual mission duration capability of five hours can be prolonged up to seven hours on customer demand. This widens the mission flexibility to inject payloads into different orbits on one flight.

Eurockot is currently offering either a classic western clamp ring separation system as baseline for the ring adapters or a Russian single point pyro attachment system, both of which are flight qualified.

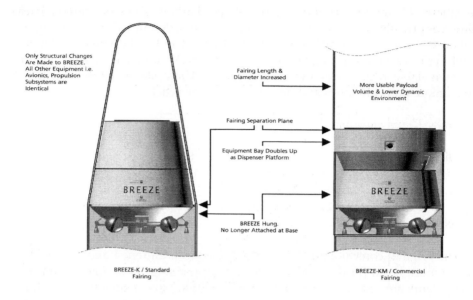

**Figure 2.** Commercial payload fairing heritage

## 4.    Performance

Rockot can be launched from both the Plesetsk Cosmodrome in Russia and the Baikonur Cosmodrome in Kazakhstan. Because both launch sites are landlocked, specific drop zones away from populated areas are reserved for the impact of separated rocket booster stages. Launch azimuths are therefore limited to those that will result in impacts in these zones. To reach orbit inclinations corresponding to other launch azimuths, Rockot trajectories can include a dogleg maneuver during the second-stage burn. If necessary, the restartable Breeze upper stage can also perform plane change maneuvers. The multiple burn capability is also used to perform the circularization burn for orbits above approximately 400 km.

From Plesetsk, the available launch azimuths are 90°, 40°, 18°-7.5°, and 345°. These correspond to inclinations of 63°, 73°, 82°-86.4°, and 97°, respectively. However, to avoid a second stage impact in foreign territorial waters, launch along the 345° azimuth must perform a dogleg during the second-stage burn, which results in injection into a 94°-inclination transfer orbit rather than the expected 97° orbit. Sun-synchronous or other retrograde orbits are reached from one of these two transfer orbits using a plane change maneuver (see Fig. 3).

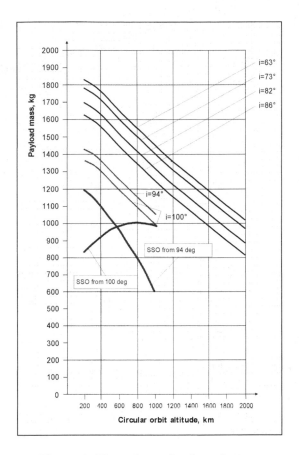

**Figure 3.** Plesetsk payload performance

From Baikonur, the restartability capability of Rockot's upper stage Breeze allows inclinations some degrees lower than 50° and higher than 53° to be reached, the exact inclination depending on payload target orbit — and mass (see Fig. 4). Rockot can also inject payloads into elliptical orbits. Payload injection into orbits which escape from the Earth or go towards the Moon is possible by using a suitable additional solid propellant kick-stage (e.g., Thiokol's STAR 37M) housed inside the fairing.

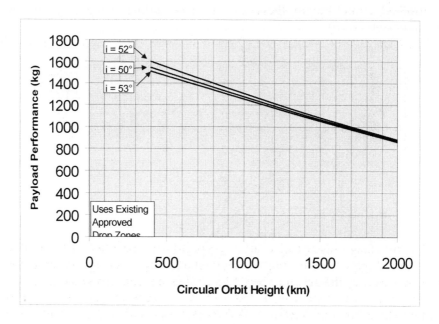

**Figure 4.** Rockot payload performance from Baikonur

## 5.    Rockot Commercial Demonstration Flight

The fully instrumented Commercial Demonstration Flight (CDF) was carried out on 16th May 2000. The CDF

- Demonstrated the operational readiness of the newly installed Plesetsk launch infrastructure for commercial operations, and

- Provided flight verification of the commercial configuration of the Rockot launch vehicle incorporating the commercial payload fairing and the Breeze-KM stage

Two identical satellite simulators, named SIMSAT-1 and SIMSAT-2, were deployed into Low Earth Orbit. The spacecraft environment was typical for Rockot commercial flights.

After the successful completion of this mission, Eurockot's Rockot Launch System was declared ready for commercial operations from Plesetsk.

## 5.1   Performance and Flight Sequence

The CDF parameters shown in Table 1 are typical for highly inclined orbits.

| Parameter | Value |
|---|---|
| Circular orbit altitude | 540 km |
| Inclination | 86.4° |
| Eccentricity | 0 |
| Right ascension of ascending node (RAAN), referred to Greenwich and fixed at launch (Ω) | 34.8° |

**Table 1.** Orbital parameters

The first stage flight ended after 136 s; before separation of the first stage, the second stage's Vernier engines were started up. Fairing deployment was performed during the second stage flight. The Breeze third stage main engine was then fired. After main engine shut down, Breeze moved into an elliptical intermediate orbit with an altitude between 153 and 540 km with an orbit inclination of 86.4°. During the following coast phase orientation of the Breeze towards the Sun was performed. Before circularization into the target orbit, Breeze was turned into the main engine firing position. After circularization at about 4500 s, Breeze was oriented into the spacecraft separation position and another propellant settling maneuver was performed using the 4 x 400 N thrusters. The two identical spacecraft, SIMSAT-1 and SIMSAT-2 were separated from Breeze keeping their orientation along the longitudinal axis. Now, Breeze attitude control was possible using only engines perpendicular to the flight plane. The upper stage was withdrawn from the spacecraft operational area. Once the spacecraft were at a safe distance from the stage, attitude control was performed using all thrusters. Breeze was then brought into position for the final main engine firing to bring the upper stage into a safe orbit of 214 km by 540 km. Flight data were down-linked to the ground control station. Fuel residuals were burnt out and pressurization gases are released from all propellant tanks.

## 5.2   Typical Launch Campaign at the Plesetsk Range

Launch Operations for Rockot are conducted by the Russian Strategic Missile Forces. Eurockot's Integration Facility (MIK) includes a state-of-the-art western standard class 100,000 cleanroom for testing the spacecraft, including a fuelling hall and CCTV monitoring systems. Also Eurockot's infrastructure includes a modern communications infrastructure enabling international communications within the Eurockot areas with hand held walkie-talkies as well as supporting fixed line phones and a local fiber optic network for communications from the launch site / pad area, processing facility and the modern remotely located launch control center (LCC). A commercial high standard hotel operated by Eurockot's parent company Khrunichev is also available. See Fig. 5 for a plan of the launch pad area.

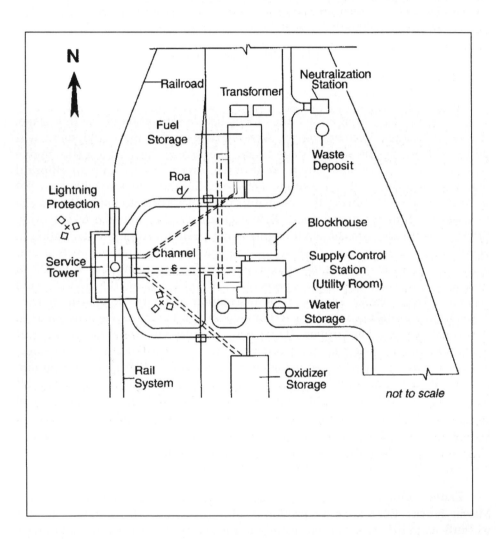

**Figure 5.** Rockot launch pad facilities

Approximately two weeks are required for spacecraft operations and many tasks can be conducted in parallel. Combined launch vehicle and payload operations take about ten days. The spacecraft are encapsulated into the fairing and attached to the Breeze upper stage. Five days before launch the booster is delivered to the launch pad in its transportation container and lifted into a vertical position. The upper part (Breeze, payload and fairing) is lifted in the service tower, fixed to the booster, and a launch transport container extension is put in place. One day before launch, the booster is fuelled through interfaces in the launch container. The service tower rolls back 10 minutes before launch. The status of the launcher, payload, pad facilities, downrange tracking and communications network is monitored throughout.

## 5.3   Rockot Radio-visibility

Downrange tracking and telemetry stations provide coverage up to Breeze first burn; further data are stored onboard and down-linked. Spacecraft deployment (5840 s) is tracked directly. The CDF radio-visibility (see Fig. 6) is typical for commercial Rockot, highly inclined ascent trajectories from Plesetsk.

## 5.4   Flight Environment during CDF

The mechanical and electrical environment on the payload will largely correspond to the conditions during future commercial launches. During ascent, the payload will experience flight time dependent quasistatic and dynamic loads in the launcher's longitudinal and lateral direction. Accelerations are initiated by the launcher itself, by mission dependent events, the environment and the selected flight path. Table 2 provides typical flight data. This will be confirmed after detailed evaluation of the measured results from the recent flight. Preliminary evaluation of the flight data indicate that the current User's guide data is conservative.

The acoustic environment under and outside the fairing will be tested during launch and throughout the ascent phase, respectively. Measurements designated "AB" and "AH" are made during flight, all the others are measured parameters on the acoustic noise field during tests in an acoustic chamber. Transducers are installed in the payload area and on the service tower.

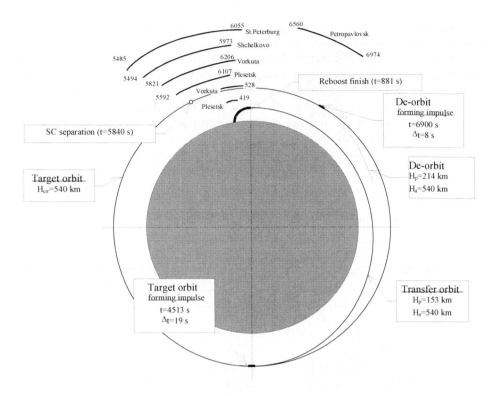

**Figure 6.** Radiovisibility during flight; the numbers at the top refer to the flight time in seconds measured from launch

| Event | Condition | Quasistatic acceleration (g) | | | | |
|---|---|---|---|---|---|---|
| | | Longitudinal (x) | | | | Lateral |
| | | Static | Dynamic | max | min | (y/z) |
| 1 | Lift-off | +1.8 | +0.5 / -1.3 | +2.3 | +0.5 | ±0.3 |
| 2 | Wind/Gust | +2.8 | 0 | +2.8 | 0 | ±1.0 |
| 3 | Max. thrust of | +7.1 | ±0.9 | +8.0 | +6.2 | ±0.3 |
| 4 | Max. g-loads of | +7.2 | ±0.9 | +8.1 | +6.3 | ±0.3 |
| 5 | Stage 1 engine | <+0.05 | +8.1 / -2.0 | +8.1 | -2.0 | ±0.3 |
| 6 | Max g-loads of | +3.0 | 0 | +3.0 | +3.0 | ±0.4 |
| 0.4 | Flight of upper | +1.6 | 0 | +1.6 | +1.6 | ±0.5 |

**Table 2.** Typical spacecraft initial design accelerations which will be confirmed by flight data evaluation

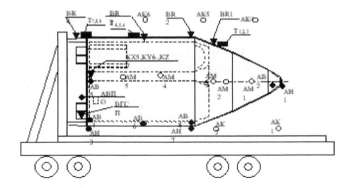

- Acoustic sensors which are using inflight and tests in acoustic chamber
- Acoustic sensors which are using only during tests in acoustic chamber

**Figure 7.** Diagram of measurements made during acoustic tests

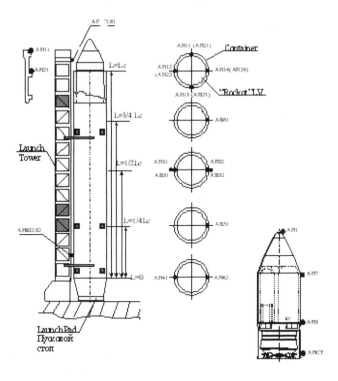

**Figure 8.** Diagram of launch acoustic measurements

## 6.     Conclusions

Following the Rockot launch vehicle's successful launch on May 16th, EUROCKOT has declared Plesetsk operational for commercial operations. EUROCKOT is thus able to offer this proven system with the following advantages:

- *Rockot* is an extremely reliable flight-proven reliable launch vehicle with a highly manoeuvrable upper stage for complex in-orbit manoeuvres to aid in multiple satellite release

- *Rockot* provides for a wide range of payload masses up to 1850 kg and multiple satellite missions into LEO and MEO orbits with a wide inclination range

- EUROCKOT offers full mission management and full mission analysis within its services:

    o   A short flexible launch campaign duration can be adjusted to the customers' needs

    o   EUROCKOT provides all services from a single source.

### References

For further information on the Rockot launch vehicle and launch services provided by Eurockot launch services, please refer to the following references given below.

1.     Eurockot Website <http://www.EUROCKOT.com>.
2.     ROCKOT User's Guide , Issue 2 rev. 1 available on web site

# Ariane 5 Evolutions: Fitting the Market Needs

P. Couillard, AEROSPATIALE MATRA Lanceurs, 66 Route de Verneuil BP3002, 78133
Les Mureaux Cedex , France

e-mail: philippe.couillard@lanceurs.aeromatra.com

**Abstract**
    Ariane 5 entered commercial service in December 1999, and Arianespace decided
in January 2000 to stop the production of Ariane 4. These two events represent a key
evolution in the European launch policy. After 20 years of being in a leading position in
the commercial market, thanks to the Ariane 1 to 4 family, the European launch
industry will rely on Ariane 5 to continue to be a key actor in the launch services arena
in the next 10 to 15 years.
    Ariane 5 is a good solution for the present conditions of the commercial space
transportation market, but it is generally acknowledged that this market will evolve
further in two key aspects :

- Stronger competition
- Increase of mass of GEO satellites.

    The European Ariane community is already coping with these challenges, through
the optimization of Ariane 5 production, and the development of new configurations of
the launcher (Ariane 5 Plus program), which will offer a progressive increase of
performance.

## 1. The Ariane Family at a Turning Point

The year 1999 has seen two important events which represent a milestone
in the history of the Ariane launcher family. After having been qualified in 1998,
Ariane 5 entered commercial services in December 1999, with the successful
launch of XMM, an X-Ray astronomy satellite for ESA. This flight was followed
by the launch, in March 2000, of two telecommunications satellites, Asiastar and
Insat 3B, in a standard commercial configuration, fully representative of future
Ariane 5 operations. Both flights were nominal, with a good injection accuracy,
and a very "quiet environment" for the payloads (acoustic, vibrations, shocks).

In January 2000 Arianespace decided to stop the production of Ariane 4
after the on-going P9.9 batch of 20 launchers. Ariane 4, the last evolution of the
initial Ariane 1 which had its maiden flight in December 1979, entered
commercial service in 1988 and was, up to now, the commercial workhorse of
Arianespace. Ariane 4 has accumulated 96 flights, with a success rate of 97%
and an on going series of 54 successive flights without failures. The quality of
the services offered by this launcher (accuracy, environment, reliability,
planning flexibility) has allowed Arianespace to occupy the leading position for
the commercial market of GEO launch services, with about a 50% share. But this
impressive background should not hide the fact that Ariane 4 has more and

93

*M. Rycroft (ed.), The Space Transportation Market: Evolution or Revolution?, 93–100.*

more difficulties to fit with the market evolution. The increase of the satellites mass (see hereafter) impedes Ariane 4 to keep on with a double launch policy, thus impacting its economical profitability. For that reason, Arianespace has decided to rely more and more on Ariane 5. The launch rate of Ariane 4 (about 10 per year today) will be progressively decreased, down to 3 in 2003, the last year of its operations.

From 2003 onwards the European launch industry will depend on the Ariane 5 launches in order to continue to be a key actor in the launch services arena in the future.

## 2.    The Market Environment for Ariane 5

### 2.1    Global Market Characteristics

The commercial market of launch services will still be dominated by the demand for telecommunications satellite. During the last 3 years, the potential development of LEO and MEO constellations has generated some hopes of a large increase of launch demand. The failure of the Iridium operation and the difficulties faced by some other constellations to raise budgets (ICO, Skybridge) have led to some moderation regarding the forecast for this segment due to these uncertainties. Therefore the core segment of the launch commercial market will remain the GEO satellites segment.

As shown in Fig. 1, the recent market analyses from Arianespace and FAA forecast a stable GEO launch demand for the coming decade, at about 25 to 30 satellites per year.

The continuous increase of satellites masses, which has been observed for the past 20 years, will go on in the future, as shown in Fig. 2. This evolution will require more powerful launchers.

### 2.2    The Competitors

The main competitors for Arianespace are Boeing and Lockheed Martin. They are preparing new families of launchers (Delta 4 and Atlas 5) which have a cost reduction target of 30 to 50% with respect to their on-going launchers. They are also controlling through joint ventures (ILS for Proton, Sea Launch for Zenit) the low cost Russian and Ukrainian launchers which can play a role in the GEO commercial market. Other competitors are the Chinese with the Long March family, who are proposing very low cost launches, but are limited by production rate and a quota policy. It is anticipated that the launch supply for the coming decade in the commercial market will be larger than the demand, making the competition harder, with a focus on the prices.

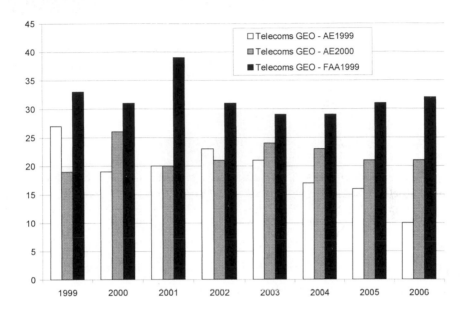

**Figure 1.** GEO satellites launch market forecast

## 2.3  *The Key Factors of Success for Ariane 5*

From the characteristics of its commercial environment, the key factors which will guarantee for Ariane 5 a similar success that Ariane 4 had are:

- To be able to reduce its launch costs to face the price competition

- To be able to satisfy the satellite's mass increase, with a prerequisite to keep a double launch policy, which is mandatory for the cost effectiveness.

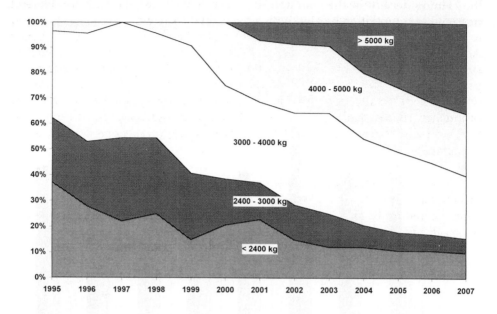

**Figure 2.** Evolution of GEO satellite masses (source Arianespace [Reference 1])

## 3.    Facing the First Challenge:  Price Competition

AEROSPATIALE MATRA Lanceurs (AML), as Industrial Architect and Prime Contractor for Ariane 5 main stages (central cryogenic stage EPC and solid boosters EAP), is fully engaged in a process of cost reduction for Ariane 5 production.

This process is mainly focused in four directions:

• EPC integration cycle reduction

• EAP integration cycle reduction

• Mission analysis and flight software cycle reduction

• Redesign to cost.

The actions taken to reduce the stage integration cycle have obvious consequences on the costs and allow the Ariane 5 launch rate to be increased up to 8 per year, which is the target ensuring the greatest economic efficiency. Actions for the EPC integration at Les Mureaux have already been implemented and are producing visible results. Regarding the EAP integration in Kourou, the process is more complicated, since the improvement implies some common operations to be elaborated with Europropulsion (responsible for the solid propulsion engine (MPS)), which in fact means an optimization of the overall MPS and EAP integration activities. Here again the improvement process has been initiated.

It should be stated that this kind of progressive improvement of stage production and integration, leading to important cost savings is not new for the Ariane community and AML in particular. Fig. 3 shows the gains obtained on Ariane 4 from batch P8 up to the last on-going batch P9.9.

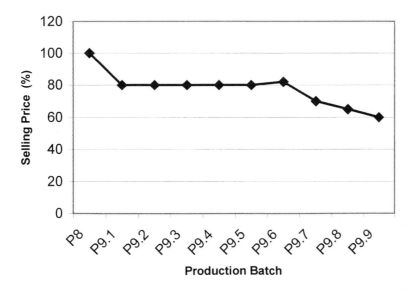

**Figure 3.** Ariane 4 cost decreases

The reduction of the mission analysis and flight software cycle is aiming not only at reducing the costs, but also at giving to Arianespace the same level of flexibility as for Ariane 4, with respect to any disturbance to the launch schedule. The year 1999 has seen numerous delays coming from the customers (satellite problems) which have imposed changes on the launch manifest, requesting new mission preparation activities, to be performed with a very short notice. This capability to react with a very high flexibility gives commercial advantages.

In addition to these actions on the optimization of the production cycles, important work has also been engaged to reduce the costs through design evolutions. This Redesign To Cost program is not aiming at changing the overall Ariane 5 architecture and technical choices, but is considering, at equipment level, some alternate solutions or technologies, which would be less expensive. In addition to improving the cost-effectiveness, key criteria of choices are to have limited impact and no risk with regard to the overall concept, and to limit the investment and the needs for new qualifications.

AML, as a key player in all these actions, is fully confident that they will be successful, and that the target for batch P2 of Ariane 5, i.e. a cost reduction of 37%, will be satisfied.

Furthermore, it is expected that the on-going European industry restructuring, such as the EADS merger, will ease and speed up the process of rationalization of Ariane 5 production.

## 4. Facing the Second Challenge: Satellite Mass Increase

The development of new configurations of Ariane 5 has been decided by ESA and is being implemented by European industry. Through a progressive improvement of its performance in GTO, Ariane 5 will be able to cope with the market constraints, while keeping a double launch policy, which is the key to its competitiveness. The major milestones of this improvement program are summarized in Table 1 and illustrated in Fig. 4.

In the frame of these new developments, AML acts as Industrial Architect and Prime Contractor for the EPC and EAP evolutions, while DASA is Prime Contractor of ESC-A development, with a strong support from AML since this stage is largely derived from the Ariane 4 Third Stage which is under AML leadership. Here again, the synergies offered by the EADS merger should ease this development.

| | GTO Performance | Availability | Improvements |
|---|---|---|---|
| Ariane 5 | 6 tonnes | 1999 | |
| Ariane 5 Versatile | 8 tonnes | 2002 | - Vulcain 2 engine (22 t additional thrust)<br>- EPC propellant increase (15 t)<br>- EAP engine (30 t additional thrust)<br>- EAP structure lightening (2 t) and propellant increase (2.4 t)<br>- multiple ignition EPS (upper storable propellant stage) |
| Ariane 5 ESC-A | 10.5 tonnes | 2001 | - same EPC and EAP changes<br>- new cryogenic upper stage ESC-A, based on HM7B engine, with 14 t propellant |
| Ariane 5 ESC-B | 12 tonnes | 2005 | - same EPC and EAP changes<br>- new cryogenic upper stage ESC-B, based on a new Vinci engine, with 24 t propellant |

**Table 1.** Evolution of Ariane 5 configurations

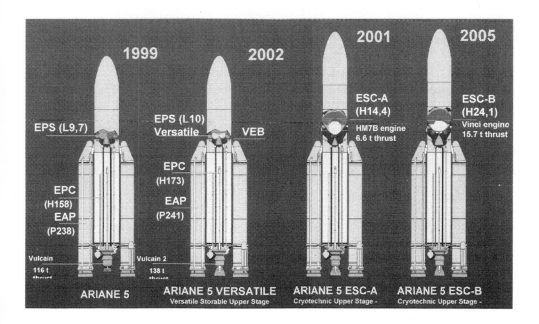

**Figure 4.** Ariane 5 configurations catalog

The introduction of the new configurations will be phased with the start of new production batches. For instance, the Ariane 5 ESC-A will enter its service with the P2 batch. This phasing will allow Arianespace to keep on track the production efficiency efforts which are implemented in the frame of a batch order, and minimize the risks associated with the introduction of a new configuration.

## 5.  Conclusions

European launch industry it at a turning point of its history. After 15 years of impressive performances, Ariane 4 will leave operational service in 2003. Its successor is Ariane 5, which already entered commercial service in 1999. European has initiated strong actions to reduce the cost and improve the performance of this launcher, allowing to face with confidence the stronger competition of the coming decade.

**References**
1.    Arianespace Website

# Potential of the Market for Space Technologies, Goods and Services in the 21st Century: A View of Russian Business

G. Khozin, A. Golovinkin, O. Pivovarov, S. Konyaev, Consulting Group "Mir", No 217 Building 1, 6 Degtyarny pereulok, Moscow, Russian Federation

e-mail: megian1@orc.ru

**Abstract**

The paper presents a long-range forecast of the development of the world market for space technology, goods and services in the 21st century, and estimates the prospects of activities in this market by Russian scientific-industrial amalgamations (NPOS) and industrial enterprises with different kinds of property. At the outset of the 21st century private business more and more actively presses for the meaningful participation of private business in the markets of space technology, goods and services. However, it cannot progress world-wide space activity unless the global space infrastructure continues to gain systematic support from the federal budgets of states.

The prospects for Russian business are analyzed. The paper proposes some new principles for the organization of transnational consortia and other commercial bodies, which will participate in the development of the world market of space technology, goods and services alongside federal agencies. Important related areas are sustainable development, using remote sensing information more widely, using space resources wisely (e.g., recycling materials from "dead" spacecraft, using resources on the Moon and Mars), global security and meteoroid/asteroid threats. Three specific examples of transnational projects are mentioned - a biosphere consortium, a material recycling plant, and an information/education center.

## 1. Introduction: Reassessment of the Space Heritage of the 20th Century

Russian business is now young, dynamic, and free from the stereotypes of the industrial age. In its desire to find its place in domestic and world markets, and in particular in the world market of space technologies, goods and services, Russian space business is ready to trespass national egotism and ambitions. It actively searches for non-traditional solutions, and at the same time uses the best traditions of the philosophy of "Russian cosmism", on the basis of which prominent Russian thinkers have put forward a long-range strategy for the exploration of the Universe for the benefit of all humankind.

The first decades of the space era were marked by a period of irreconcilable military and political rivalry and ideological confrontation, the Cold War. The strongest rationale which motivated the actions of the first "space powers" — the USSR and the USA — was the desire to strengthen their positions in rockets and nuclear confrontation for military domination on planet Earth. This was the reason why the humanistic designs of K. Tsiolkovsky, who believed that the exploration of space will provide humankind with "mountains of bread ... the abyss of power", as well as the hopes of A. Einstein that space

*M. Rycroft (ed.), The Space Transportation Market: Evolution or Revolution?*, 101–110.

will become a realm of creative and exceptionally peaceful activity, were moved into the background of the motives for space activity. In the early history of the space age, space activity contributed to the separation of nations and peoples, and entertained nationalistic ambitions rather then served the ideals of holistic and harmonic civilization.

Pragmatic as it is, Russian space business respects the high ideals of civilization, which, in essence, are inseparable from the present and future activities of humankind both on Earth and in outer space. Russian business considers space to be a natural environment, and where transnational cooperation and honest competition must develop, where integration should dominate. This is the only possible way of avoiding stagnation both in the interaction of man and the biosphere and in the exploration and practical use of outer space. Only when motivated by ensuring survival and comprehensive security of the world community at large, and by optimization of the economic development of states and regions world-wide, may space activity have a positive impact on the globalization of international relations and the world economy.

Russia has at its disposal vast resources for the successful development of space activity, taking due account of the changing conditions at home and on the world arena. The unique potential of Russian space technology, the priceless experience of operation of space assets for rather a long period of time as well as of participation in international cooperation in the exploration and practical uses of outer space, combined with an industrial infrastructure and professionals of the highest expertise, serve as a sound foundation for a considerable improvement of Russia's position in world-wide space activity by the joint efforts of federal agencies and private business.

Among the promising areas of space activity where the Russian Federation may gain real success, we may pick out the production in space of materials and products with unique properties, which cannot be made under conditions of the Earth's gravity, the improvement of technologies for non-space branches of the economy using the latest achievements of the space program and offering a wide range of commercial services to customers using space systems.

The last several years indicate that Russian and foreign customers, users of services, partners and investors are not well aware of the nature and the real possibilities of the Russian space program. The reasons for this are interagency barriers, and the lack of appropriate experience and qualified personnel in the field of space commercialization in Russian agencies and organizations as well as the low rate of transfer of products of the Russian space program to internal and world markets.

Russian space business has its origins in the depths of the missile and space branches of the Soviet economy, which were strictly controlled and managed by the command and control system of the government of the Soviet Union. At present, Russian space business is reinforced by the brand new (for Russians) socio-economic structures and organizations, which show interest in the implementation of private projects aimed at the development of specific types of space technology, and rendering services by means of space systems, as well as in consulting and marketing of space technologies, goods and services, etc..

Representatives of Russian space businesses do their best to maintain mutually beneficial contacts with the Russian Aerospace Agency (Rosaviakosm) and other federal agencies participating in the implementation of the national space program of the Russian Federation. They pursue promising new areas of the economically efficient use of space technology beyond the limits of the federal space projects which are strictly controlled by the government. It is exactly the Russian space business which is deeply interested in the positive development of the global space infrastructure as well as in its liberation from the pragmatic interests of individual states and narrow transnational groups.

To increase the role of Russian space business in space activity, measures are taken to provide more efficient consulting assistance and services to potential customers, users of services and investors. Calculations show that at present the annual volume of investments attracted by the aerospace sector of the Russian economy amounts to about US $ 800 million. Further improvement of consulting support for participants of the space activity, aimed primarily at raising the efficiency of marketing of the products of the Russian space program on internal and world markets, will make it possible not only to increase the volume of investments into the aerospace sector but also to create powerful incentives for progress in other high-technology sectors of the Russian economy.

The politicians, scientists and representatives of the business communities concerned identify the following issues, whose solution will provide for the steady development, with the appropriate involvement of private business, of the global space infrastructure under conditions of the globalization of international relations and the world economy, and thus would contribute to sustainable development of the planet and its civilizations:

• Implementing a realistic and interdisciplinary long-range forecast of the evolution of global space infrastructure for the next 100 years or so. This task may be assigned to an international group of experts, which may be established under the aegis of the International Space University. Such a forecast will be the logical follow-on of the ideas, put forward by such Russian thinkers as N. Fedorov, H. Kibalchich, K. Tsiolkovsky, Yu. Kondratiyk, F. Tsander and S. Korolev, and the concepts proposed by such European engineers and scientists as H. Oberth, E. Sanger, R. Esnault-Pelterie and H. Noordung as well as by American researchers G. O'Neil, H. Kahn, C. Ehricke and R. Jones

- Deepening and expanding of sustainable direct and feedback interactions of the global space infrastructure with the current economic and socio-political activities of individual nations, regions and, indeed, the entire world community. A solution of this problem will require an expansion of the sphere of the practical utilization of the space infrastructure for the benefit of the entire population of the planet

- Perfecting the international cooperation, coordination and integration of efforts of all the participants of space activities. For success in this area it is necessary to decrease the number of space projects aimed at military-political rivalry, to withdraw space systems from being under the unilateral control of military establishments, and to agree with the rationale of the incorporation of space business into the ranks of the primary actors of world-wide space activities.

## 2.    Potential of Private Space Business as a Participant of Space Activities

As space business gains access to space technologies, goods and services it may become a prime mover for the widest possible commercialization of space activities. The emergence of space business as a major participant of space activities will become ever more vigorous, and will widen its relations with customers, investors and the public at large. Private business, including Russian business, is vitally interested in establishing reliable and efficient interrelations of the global space infrastructure with the actual needs and demands as well as with the economic possibilities of the world community on global, regional, national and local levels. In a search for the ways and means of providing a viable space infrastructure, space business may overcome the inertia of space activity dominated by federal management.

At the outset of the 21st century, a meaningful share of the appropriations for world-wide space activity comes from international space communications consortia, where private business plays ever more important roles. The share contributed by private business to the improvement of national, regional and global space systems for communications, remote sensing, weather forecasting, navigation, etc., will grow in the foreseeable future. As this favorable trend continues, the access of customers and users to new space technologies will increase. Some of the new space technologies will be independently developed and operated by the organizations of private business. While doing this, private business will maintain and strengthen its interaction with space enterprises controlled by the government. Thus, it will stimulate the process of commercialization of space activities and society's increase of space goods and services.

3.    **Major Goals and Directions for the Development of the Global Space Infrastructure in the 21st Century**

Following the best traditions and relying on the creative legacy of the forecasts of space activity which have been made in the 20th century, the evolution of the global space infrastructure in the 21st century should be based on some novel principles, the most important of which are:

• Confrontation-oriented geopolitical concepts and egotistic national interests, which may be used as a rationale for space activity, are counter productive

• The space environment cannot be regarded as the property of individual states or international blocks

• Space activities may be the only realistic means of ensuring the survival of human civilization in the face of the threat of collision with a meteoroid or asteroid

• Disengaged nuclear weapons may be stored in low Earth orbit or on the Moon with the possibility of being used to protect civilization from incoming meteoroids and asteroids

• A critical attitude to Earth-based nuclear power generation should not become on obstacle to the development of nuclear engines for orbit-to-orbit and interplanetary tugs

• Space debris is a significant hazard to future space activities

• Space activities remain a source of human inspiration.

The steady expansion of space activities should assist development of the Earth and its sustainable civilization. The following are specific features of this model:

- Stabilize the biosphere by means of balancing the regeneration of biota and anthropogenic impacts

- Limit the growth of the Earth's population so as not to overload the biosphere

- Satisfy the reasonable needs of each citizen in a eco-friendly way

- Improve the intellectual and spiritual activity of humans using global information networks.

The global space infrastructure, which will be developed and improved step by step, will:

- Provide civilization with wider possibilities for the exploitation of the resources of the Moon and planets

- Supply the world community with ever more comprehensive remote sensing data, and

- Make such information supporting the needs of civilization more reliable.

4.   **Priority Tasks**

Top priority goals may be to:

- Create a global security system to protect the planet from meteoroid and asteroid threats; such a system will have ground-based and orbital infrastructure to collect and process information on them to prevent their collision with the Earth, and, if necessary, to "attack" them with nuclear weapons

- Establish permanent bases on the Moon (around 2030) and on Mars (around 2080) for research and production, exploiting the "local" natural resources, as launching sites for the further exploration and practical use of the planets of the solar system, for maintaining elements of the global security system, and for creating artificial biospheres

- Establish a permanent base at the geostationary orbit to reuse the resources of spacecraft which have exceeded their active lifetime

- Provide a space segment for a few new transnational associations, which could unite the efforts of federal agencies, international organizations, transnational corporations and private businesses from individual states to create space systems designed first and foremost for the benefit of humanity as a whole; private businesses may, for example, supply fuel cells and other on-board energy sources

- Design an international organization for using information derived from space systems (e.g., for global ecological monitoring, remote sensing in the interests of mining industries, industrial production, agriculture, services, etc.)

- Continue research and development projects aimed at the creation of advanced and highly competitive space technologies, and involving private business in such areas as space processing of disease-fighting drugs and pharmacological preparations, the growth of super-clean crystals and the production of new composite materials.

## 5. Mission-oriented Programs to be Implemented with the Participation of Private Business

Adverse conditions affecting the Russian national space program during the recent period of radical political and socio-economic reforms are responsible for the considerable reduction of appropriations for space activity, for decreasing the annual number of spacecraft launches as well as for narrowing the spectrum of advanced research and development programs. However, among other factors due to the progress in space commercialization and an increased interest of private business to participate in space projects, Russian scientists and businessmen have put forward some competitive projects. Such projects may be successfully implemented by the joint efforts of the federal space agencies of different states, international organizations, transnational corporations and private business. Some details of three such projects are given.

The Biosphere consortium is a global network of regional centers for the protection and rational utilization of the resources of the biosphere. These centers will closely cooperate with each other, primarily by means of providing access to space assets for ecological monitoring, remote sensing, and collecting, processing and disseminating different kinds of information. The major elements of the biosphere consortium will be:

- A global biosphere information center, comprising an orbital center for the collection of information about the state of the biosphere and the ground control center with two sub-centers in western and eastern hemispheres

- A satellite network of biosphere information, which will unite the operational national and regional satellite systems (Landsat, SPOT, Meteor, etc.), receive appropriate information from manned space stations (Mir, International Space Station, etc.) and cooperate with future spacecraft designed to collect biosphere information

- Continental and sub-continental biosphere information centers

- Regional centers for the protection and rational utilization of biospheric resources.

Since the primary mission of the biosphere consortium will be provision of a wide and constantly expanding spectrum of services, its customers will include numerous federal and regional departments and agencies, national and transnational corporations, organizations of medium and small businesses. Among its clients and users of its services may be presidential administrations and offices of the heads of governments, national legislative bodies, federal departments and agencies responsible for national security, industrial and agricultural production, mining, utilization of energy and other natural resources, transport systems, control of the quality of the environment and weather conditions, health and education of the population, and dissemination of information. Apart from this, the biosphere consortium will provide valuable services and unique information to regional authorities, local administrations and private business engaged in insurance, finance and banking and public order enforcement.

The "Kosmut" (space utilization) Project is to reuse the construction materials and equipment of Mir after all possible modes of its active life have expired. It may be possible to utilize in a similar way materials from other "dead" spacecraft as well as some space debris.

The materials (aluminum, titanium-magnesium alloys, silicon, other valuable metals and chemical elements as well as plastics) and equipment of Mir, whose total weight is about 135 tonnes would be reprocessed into fine-grained materials for use in the rocket engines of orbital tugs and other means of transorbital flight in combination with an oxidizer delivered from Earth.

In its present form, project "Kosmut" has an estimated capital cost of US $ 2.7 billion. The project has three priority goals:

- Development of methods to produce an efficient fuel mixture out of construction materials used in the station with minimal additions delivered from Earth

- Design and construction of an orbital tug to transfer construction elements and new spacecraft to higher orbits

- Establishment of a modular orbital structure to dismantle spacecraft and process some parts of them into fine-grained fuel substances.

To ensure favorable conditions for the operation of Mir in this new role it may be necessary to put it into a higher orbit. Initial estimates indicate that with a daily average of processing 100 kg of construction materials (at a cost of few US $ thousand per kg) the station may produce 36 tonnes of fuel mixture per year.

The experience gained may be used later for the International Space Station, which may become a base for the establishment of a utilization center in geostationary orbit.

The Information and Education Center (Education Department "Mir") uses new educational technologies to become the central link of a global system of university centers of innovative education, based on an interdisciplinary model of sustainable civilization. Lecture courses, seminars, discussions and individual lessons, which form the curricula of the educational department "Mir", will be carried out by means of contacts with users through the Internet and other global information networks as well as by transmission to mobile TV reception stations, which are at educational establishments on the ground.

The customers of the education department "Mir" may receive information directly from the orbital complex as well as communicate with cosmonaut (astronaut)-lecturers working in space (in the initial stage aboard the space station Mir or, later, on the International Space Station). Lectures and seminars organized by the education department "Mir" will be delivered in Russian and in English and may be simultaneously translated into other languages.

The approximate cost of a lecture (one and a half hours) may be up to US $ 100 thousand, and may bring in an income of about US $ 200 thousand. The income may grow if special audio-cassettes, compact discs and special educational films were to be produced.

## 6. Conclusion

In conclusion, it should pointed out that this paper covers only the most promising (from the viewpoint of Russian experts and businessmen) projects and possible areas for space commercialization. Russian space business is open to wide contacts and joint initiatives which will foster both greater efficiency and competitiveness of the global space infrastructure. Together, we should pioneer the commercial space frontier.

# Report on Panel Discussion 2:

# Meeting the Needs of Future Markets — Launch Service Providers' Perspectives

O. Angerer, G. Bolognese, R. Locantore, International Space University, Strasbourg Central Campus, Parc d'Innovation, Boulevard Gonthier d'Andernach, 67400 Illkirch-Graffenstaden, France

e-mail: angerer@mss.isunet.edu, bolognese@mss.isunet.edu, locantore@mss.isunet.edu

**Panel Chair : G. Laslandes, CNES, France**

**Panel Members:**

**P. Bonguet,** Starsem, France
**D. Buck,** United States Air Force, USA
**P. Couillard,** Aerospatiale-Matra Lanceurs, France
**P. Freeborn,** Eurockot, Germany
**R. Gao,** China Great Wall Industry Corporation, China
**J. Honeycutt,** Lockheed Martin Space Operations, USA
**G. Khozin,** Consulting Group "Mir", Russia
**B. Parkinson,** ASTRIUM (Matra Marconi Space), UK
**P. Rudloff,** Arianespace, France
**E. Stallmer,** The Space Transportation Association, USA

For the second discussion panel, representatives and consultants of various launch service providers presented and discussed their views on future launch markets and how the future demands might be met.

The trend towards larger, multi-national corporate entities was discussed. Rationalisation of launch services and operations is the driving force towards consolidation. As well, there is greater co-operation among certain companies, such as Arianespace, Eurockot and Starsem, to complement each other's product and to reduce duplication. Such integration would present launch service providers with greater flexibility in meeting customer needs and ensuring competitiveness. For example, the Arianespace group concentrates its efforts on offering a harmonised family of launchers for differing market needs. However, political considerations are influencing further harmonisation and standardisation. Europe wishes to maintain independence in the launch service market, as technical co-operation with US companies may be hindered by United States Government-imposed restrictions.

*M. Rycroft (ed.), The Space Transportation Market: Evolution or Revolution?*, 111–113.
© 2000 *Kluwer Academic Publishers. Printed in the Netherlands.*

Currently, space industry forecasts project a drop in demand for their launch services. Recent notable failures in the constellation segment of the space industry have resulted in greater customer perception of market risk. When questioned about how launcher demand can be increased in the future, the subject of space tourism was discussed as a possibility. The space tourism market was seen as having a revolutionary effect on the launch industry. However, several panelists remarked that current launcher companies would not lead the development of suitable launch vehicles; they would only invest in such projects if and when the space tourism market develops.

The question was raised as to whether launch vehicles developed by smaller players such as Brazil and India would be able to compete with established launch service providers. It was noted that such market entrants would be able to compete with US launch service providers, as the US providers are traditionally more expensive. On the other hand, they would not be able to compete against Russian and Russian-derived launch vehicles such as Eurockot, as Brazil and India would not be able to match their less expensive launch prices. So far, these nations and others, such as China, have been serving domestic launch needs since providing services on the international scene has been hindered by political motives such as ITAR restrictions. Even the launch service market leader, Arianespace, wishes to reduce government-imposed restrictions, and has asked the US to perform an independent security evaluation at its Kourou launch site, in the hope of capturing a greater share of the US launch market. Such restrictions put the US launch customers, the satellite manufacturing and service industry, at a disadvantage since they are forced to purchase more expensive US launches.

A growing concern is the diminishing importance of space in the public mind. In order to better compete with other social responsibilities for scarce government resources, it was stated that some form of a civilian National Space Council, separate from NASA, would be in a better position to put space at the forefront of public awareness and argue for increased resources. On the national security side, however, a hypothetical 'Space Command', separate from the United States Air Force, was dismissed. The USAF wishes to change from providing launch services to purchasing launches for its future military contracts from other providers, possibly even from non-US companies.

As environmental concerns are always an issue, the question of the effect of increasing number of launches on local populations and the environment was discussed. In the case of Arianespace, it was stated that CNES has always imposed restrictions regarding launch trajectories, so as to minimise potential dangers to the local population in Kourou. As well, at a yearly launch rate of 10 per year, the toxic emissions from Ariane launches have a minimal impact on the environment.

# Session 3

# Reusables and Expendables in the Future Launch Market: What will be the (Right) Mix? And Where is Technology Heading?

Session Chair:

**S. Nomura,** NASDA, Japan

# Reusables vs. Expendables:
# to Low Earth Orbit (LEO) and Beyond

**W. Fletcher,** Expendable Launch Vehicles and Payload Carriers Programs Office, National Aeronautics and Space Administration, Mail Code VA, Kennedy Space Center, FL 32899, USA

e-mail: William.Fletcher-1@ksc.nasa.gov

**Abstract**

This paper presents an overview of some of the factors that affect the mix of Reusable versus Expendable vehicles. In developing this overview, a qualitative look at the launch market requirements will be performed. The launch market requirements are broken down into four orbital markets or flight regimes: surface to low Earth orbit, orbital operations near Earth, operations at a distance from Earth and return, and deep space probes. The current need and future use of Reusable versus Expendable vehicles in these four flight regimes will be addressed and some ideas given on the infrastructure necessary to support viable space operations.

## 1.  Space Transportation — Evolution or Revolution?

Basic rocket technology has slowly evolved since the Chinese first invented gunpowder in the eighth century and launched projectiles out of bamboo tubes. Only thirty years ago, some textbooks still taught that single stage to orbit (SSTO) was not possible given the propellants and mass fractions available at that time. Today, lighter composite materials and densified propellants make SSTO possible. And future improvements in launch ranges, and the use of space assets for launch tracking, will yield not only lighter vehicles but also improved vehicle performance. But short of a truly revolutionary, technological breakthrough in the area of propulsion, we are likely to continue doing "business as usual".

Additionally, until now we have had no long-term vision for human presence in space. Apollo was a stepping stone that took the United States to the Moon. But we have never been back. Mir was a scientific outpost, but it never achieved any operational capability to service other space vehicles, or support a space infrastructure. The International Space Station is considered to be the next step along the road, but its ability to support a truly commercial infrastructure has not yet been demonstrated.

Finally, today's mix of Reusable and Expendable Launch Vehicles is an expensive one. Launch costs per kg are some tens of thousands of dollars. And although current reliability projections in the Expendable Launch Vehicle (ELV) industry are for more than 95%, that rate will not suffice for future industry markets [Reference 1].

*M. Rycroft (ed.), The Space Transportation Market: Evolution or Revolution?*, 117–125.
© 2000 *Kluwer Academic Publishers. Printed in the Netherlands.*

How then can we achieve that revolutionary state that will make access to space easier and more affordable for everyone? Let's look for a moment at some other transportation industries. The automobile industry was in its infancy when Henry Ford revolutionized it through his concept of the automobile assembly line. The assembly line itself did not bring any new or great technology into play, but instead developed a new process of manufacturing cars that made them cheaper to build and affordable for nearly everyone. Ford had a vision that things could be done differently, and he had the drive to carry through with his vision. The aircraft industry was also in its infancy and struggling when Bill Boeing of The Boeing Company had the vision of developing markets for commercial air travel. The resulting new market demands fueled a revolution in aircraft technology across the industry and brought prices down, making air travel increasingly affordable for everyone.

Taking our cue from these industries then, the key to building a thriving space transportation industry is two-fold. First, we will need to take a good look at improving not only the efficiency of our launch vehicles, but our overall launch process efficiency as well. Second, we will need to develop space transportation markets that will enable the average person to make use of space assets and products. We must encourage the attitude that space is just another place to do business. We must come to the place where space is no longer a technical marvel, but an indispensable service [Reference 2].

## 2.    Launch Market Development

The two main factors that will encourage the development of launch markets are launch cost and reliability. The drive today is to bring down launch costs to about US $ 2,000 per kg or less, through the development of reusable launch vehicles (RLV's) [Reference 3]. Today, there are some 20 sub-orbital and orbital reusable vehicles in development [Reference 4]. But these will not all make it to full scale production. Investment capital is hard to come by because of the lack of a thriving launch market. And given development costs of US $ 1 billion plus, if one supplier could capture half of today's launch market of approximately 120 satellite launches per year [References 5, 6, 7], then over a ten year period (600 launches), this equates to approximately US $ 1.7 million per launch just to recover development costs, not counting inflation and interest charges. For the payload capacity for the vehicles under consideration (3,000 to 9,000 kg) this equates to a minimum of US $ 200 per kg. This could have significant impact on the goal of $ 2000 per kg.

The other issue is reliability. Depending on whom you talk to, the current reliability projections in the ELV Industry are about 95% [Reference 8]. Even a

1% failure rate for aircraft operations at the Orlando International Airport would lead to a crash every 2.4 hours. With man-rated space vehicles, even a 99% reliability factor is not sufficient. We must also address the issue of vehicle certification. What will the criteria for certification be, and who will set those certification requirements for man-rated vehicles? We will also need to consider the cost of meeting those certification requirements, which in the aircraft industry usually runs to ten times the development costs [Reference 9].

## 3.    The Right Mix

So what is the right mix? That question cannot be answered without understanding several things: first of all, what vehicles are available or can potentially be made available; second, what are the launch markets, and third, what is the best vehicle to satisfy the different orbital markets?

### 3.1  Availability

In today's environment, we probably have an optimum utilization of the mix of available vehicles. The Space Shuttle, which was designed as a semi-reusable vehicle to carry cargo and passengers to LEO, is clearly the best choice to continue that role in servicing the International Space Station and carrying certain critical payloads into orbit. Expendables carry all other payloads today.

It is generally assumed the industry must develop RLV's to achieve a faster turn around, cheaper operating costs, and improved safety margins. But today's RLV's are limited. The Space Shuttle, for instance, is not a truly reusable vehicle, and has the limitation of only being able to achieve low Earth orbit. Several emerging companies hope to be able to deliver a truly 100% reusable vehicle. But their current performance will be limited to sub-orbital missions or to low Earth orbit. So an expendable upper stage will be needed for any missions that go beyond low Earth orbit. The bottom line with today's and the near future's (5-10 years) technology  is that we will have to continue using a mixture of reusable and expendable  hardware in order to fulfill many of today's missions.

### 3.2  Launch Markets

The world launch market is generally conceived to be divided into three major customer bases: commercial, government-led science and research (i.e., NASA, ESA), and military. A qualitative breakdown of the launch forecast for these customer bases is shown in Table 1.

| Commercial | Government Led Science and Research | Military |
|---|---|---|
| Communications<br>•   LEO, MEO,GSO | Earth Sensing | Earth Sensing |
| Tourism<br>•   Sub-orbital<br>•   LEO (Short, Long Stay)<br>•   Moon, Planets | Human Exploration<br>•   Earth Orbit<br>•   Earth Vicinity<br>•   Planetary<br>•   Deep Space | Navigation |
| Mining<br>•   Moon,Asteroids,Planets | Science Observation<br>•   Earth Orbit<br>•   Earth Vicinity | Communication |
| Manufacturing Applications<br>•   Microgravity (LEO) | Communications Satellites (GSO) | |
| | Deep Space/Planetary Probes | |

**Table 1.** Launch forecast requirements

Much of the projected launch market has not yet been realized or fully explored. Space tourism is one area that has been addressed and some attempt made to quantify the potential market. NASA and the Space Transportation Association conducted a study in 1996 titled "General Public Space Travel and Tourism" [Reference 10]. A brief summary of some of the findings indicate that out of a US $ 400 billion tourism industry, tourism at space related activities (space camps, museums and government R&D centers) draws 10 million people per year and generates US $ 1 billion in revenues. Surveys indicate tens of millions of "average" US adults can envision themselves taking a trip into space. If ticket prices can be brought down to well below US $ 50,000, then a market of up to 500,000 space trip passengers a year would open up. This translates into a potential US $ 10-20 billion per year space tourism business.

It has been suggested by this and other reports that space tourism  should be developed incrementally. For example, early space tourism would begin with sub-orbital flights. Orbital trips could follow. As the market developed, on-orbit stays (orbiting hotels) could be  developed. Finally, resort packages could be developed. Who knows? We could be spending our holidays in a kind of orbiting "dude ranch," with day treks to the Moon:  it's not inconceivable!

A market that has not been explored is made up of the university and research communities. As indicated in Fig. 1, they make up a major portion of the NASA ELV Program customer base. Today, they generally must acquire government grants, or have a partnership with a government organization such as NASA. But, with a decreased cost of getting into space, these communities could become a substantial independent customer base. Their payloads tend to be small, however, and the challenge will be to develop methods of deploying multiple small payloads from one launcher with varying orbit requirements. Other industries that have either expressed some interest or are in-work today include advertising, waste disposal, space burial, and utilities services.

*Updated 1/14/00*

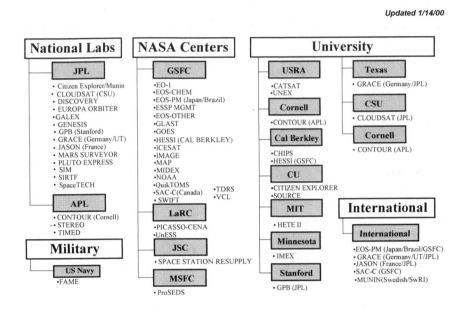

**Figure 1.** Some satellite missions in the NASA ELV program customer base

As we have already noted, most of the RLV's currently being proposed have fairly limited lift capacity. So the idea of on-orbit assembly needs to be explored for projects such as large communications satellites, or exploration of the outer planets. And no discussion of the space market would be complete without some discussion of a Moon base. Revenue from mining, manufacturing or space related research would probably make a Moon base self-sustaining. Other applications include: (1) space operations center and fueling station for space based vehicles, (2) assembly and training point for manned planetary missions, (3) planetary defense system against comets or asteroids, and (4) use of the dark side of the Moon as a "quiet" laboratory for exploration of the Universe. Use of the Moon as a spaceport for launches to the planets or deep space is particularly advantageous. The low gravity and lack of atmosphere could lend the Moon to ground-based acceleration systems such as the magnetic-levitation system which NASA is currently studying, or to a catapult system powered by steam or high pressure gases such as used on  aircraft carriers.

*3.3   Orbital Market Requirements*

Finally, we need to look at the various launch markets in terms of their orbital requirements:  LEO, Near Earth Vicinity, Distant from Earth and Return, and Deep Space. These four orbital markets (or flight regimes) were selected for two reasons. First, there is a natural distribution of the launch requirements around the four points (as illustrated in Table 2). Second, these four orbital markets require different characteristics in the launch vehicles used (as illustrated in Table 3).

| Orbital Market | Launch Market |
|---|---|
| Low Earth Orbit (LEO) | Communications at LEO<br>Tourism Sub-orbital and LEO<br>Manufacturing Applications – Microgravity<br>Human Exploration<br>Science Observation from Earth Orbit |
| Earth Vicinity | Communications at MEO and GSO<br>Tourism<br>Earth Sensing<br>Human Exploration<br>Science Observation<br>Navigation |
| Distant from the Earth and Return | Tourism of the Moon and Planets<br>Mining of Moon, Planets, and Asteroids<br>Human Exploration – Planetary |
| Deep Space | Deep Space Probes |

**Table 2.** Launch market forecast by orbital market

Each orbital market (or flight regime) has a certain set of requirements and characteristics that makes it unique. One must think in terms of these requirements when designing vehicles that will operate in each of these orbital markets. For example, the surface to LEO vehicle obviously has to be a vehicle that can operate in an aerodynamic environment as well as the beginnings of a space environment. Over the next five to ten years as the markets begin to open up, a shift to RLV's in this orbital market is clearly indicated.

Maintenance and serviceability is another consideration.  The idea of space tugs operating near Earth, moving people and cargo from LEO to other points becomes viable. Along with the movement of cargo in space, the concept of "tow trucks" to repair or bring in for repair damaged satellites or vehicles becomes a very real possibility. The design of a vehicle operating near the Earth can take into account repairability features that other vehicles cannot.

| Orbital Market | Launch Vehicle Characteristics |
|---|---|
| Low Earth Orbit (LEO) | Controls for both aerodynamic and orbital flight<br>Thermal protection for both aerodynamic and space regime<br>Guidance required for very short time periods (minutes to few hours)<br>Able to tolerate Shock Loads and G-loads from launch<br>Ground maintainable, reusable<br>In-flight anomaly prevention requires redundant systems |
| Earth Vicinity | Controls for orbital flight<br>Thermal protection for space regime<br>Guidance required for short time periods (hours to a few days)<br>Reusable and serviceable given space infrastructure<br>Good chance to recover from in-flight anomaly (send out tow-truck) |
| Distant from the Earth and Return | Controls for orbital flight<br>Thermal protection for space regime<br>Guidance required for long time periods (months to years)<br>Requires redundant and autonomous systems unless accompanied by manned crew<br>Leaves some expendable parts behind on return |
| Deep Space | Controls for orbital flight<br>Thermal protection for space regime<br>Guidance required for long time periods (months to years)<br>Not serviceable and so must be redundant and autonomous |

**Table 3.** Launch vehicle characteristics vs. flight regime

In the years following an initial push into LEO will come the opportunity truly to revolutionize space and the launch vehicles. There will need to be a renewed commitment to true manned space operations. Space transportation then becomes more like transportation here on Earth: different forms of transport will be used to move cargo and people from one point to another. Consider how some of us got to this conference: probably by automobile or other ground conveyance to the airport, and then by airplane to another airport, and then by ground conveyance to our final destination. This is the way space transportation should be handled, with a vehicle appropriate to the orbital market in which it has to operate. By the time we reach this point, nearly all launch vehicles will be reusable except those for deep space probes, which will not return and which are expendable by their very nature.

## 4.     Summary

Enough attention cannot be paid to expanding markets and to the effects which they will have on the ability of the RLV's to survive. In the current market, the civil government and military cry out for cheaper access to space, but the work and applications which they represent are too important not to continue to pay. Current profit margins for the communications satellite operators at GSO will assure their continued business. The drivers for change that will truly knock down prices will come initially from the users of LEO. The LEO telecommunications satellite market will need the price break. Their projected profit margins are tight, and the cost of launches to make up their constellations is a significant driver. Other markets in LEO similarly will never get off the ground without the price and reliability breaks that RLV's can potentially offer.

In reviewing the best mix of ELV's and RLV's, one has to consider today's available vehicles and market drivers. Today's mix is an expensive one, both with the current ELV's and the Space Shuttle. Large capital investments will be required to improve the situation and achieve lower cost and more routine access to space. In looking at improving the capability, the economic law of supply and demand has to be considered, and the two have to be considered hand-in-hand. More demand is required to drive the requirements for new capability and lead to cost-effective RLV's. Thus we should not just be looking at new launch vehicles, but we need also to be generating new markets for these vehicles at the same time. New markets lead to higher launch rates, which lead to higher profit motive, and thus to the new investment capital that is so badly needed to get the Space Transportation Industry moving.

### References

1.     Wade, M.: *Mark Wade's Encyclopedia Astronautica*, <http://www.friends-partners.org/~mwade/spaceflt.htm>. April 21, 2000
2.     Space Publications in collaboration with International Space Business Council: *State of the Space Industry, 1999*, "The Market for Consumer & Business Services is Driving the Industry", p. 14, 1999
3.     Odenwald, S./Raytheon STX: *NASA IMAGE Space Science Questions and Answers*, <http://image.gsfc.nasa.gov/poetry//ask/a11819.html>.     IMAGE     Satellite Program, NASA, April 6, 2000
4.     Space Future.com: *Vehicle Designs – Current Projects*, <http://www.spacefuture.com/vehicles/designs.shtml#SPACEACESS>. April 21, 2000
5.     Stauf, G.: A Case for RLV's – Why Destroy Expensive Spaceships, *Launchspace*, Vol. 5, Number 2, p. 34, 2000
6.     Space Publications in collaboration with International Space Business Council: *State of the Space Industry, 1999* , Launch Vehicle Market, p. 33 , 1999

7.  Caceres, M.:  *Aviation Week & Space Technology Online: Launch Vehicles: Steady Growth*,  <http://www.awgnet.com/aviation/sourcebook/99launch.html>.  April 6, 2000

8.  Associate Administrator for Commercial Space Transportation (AST): *Commercial Space Transportation: 1999 Year in Review,* January 2000

9.  Foust, J.: *Barriers to Space Tourism,*
    <http://www.spaceviews.com/1999/07/article1a.html>.  SpaceViews News, April 6, 2000

10. National Aeronautics and Space Administration: *General Public Space Travel and Tourism – Volume 1 Executive Summary,* March 1998

# Technology Experience with the H-II ELV and Its Evolution

**T. Ito**, National Space Development Agency of Japan, 2-4-1 Hamamatsu-cho, Minato-ku, Tokyo 105-8060, Japan

e-mail: ito.tetsuichi@nasda.go.jp

### Abstract

This paper presents technological features of the H-IIA launcher which evolved from the H-II ELV, H-II flight anomalies, and corrective actions for the H-IIA development program. Major ELV subjects such as robustness, reliability and cost are now being improved in the H-IIA program. The evolving ideas are incorporated not only at the components level but also at the systems level.

H-II ELVs have flown successfully five times since the first flight in 1994; then, however, two successive flights ended in failure due to an engine being premature shut off. Failure investigation efforts revealed that the root cause was related to manufacturing quality control and design conditions in both cases. Corrective actions for the H-IIA program are the addition of critical development tests and a flight test, and simplification of development works including cancellation of the H-II program, besides counter measures such as the improvement of manufacturing quality control and the execution of specific detailed tests of the engine design conditions.

## 1. Introduction

The H-II rocket, a two stage expendable launch vehicle (ELV) capable of delivering 4 tonnes to geosynchronous transfer orbit (GTO) was developed by the National Space Development Agency of Japan (NASDA) for ten years from the mid-1980s utilizing indigenous technologies throughout. It launched satellites successfully five times after its inaugural flight in February 1994. It seemed that H-II ELV technology had been practically established.

Then, in the mid 1990s the H-IIA program was started to meet diversifying launch demands, to improve robustness in production, and to reduce recurrent costs with improving operational reliability. Toward the final development phase of the II-IIA, however, the sixth H-II flight (H-II F5) and the seventh (H-II F8) ended in successive failures, in February 1998 and November 1999. Both failures were caused by premature engine shut down, of the second stage and the first stage, respectively. During intensive failure investigation efforts for the H-II F8 failure, a comprehensive review of the H-II and H-IIA programs was conducted and significant corrective actions in technology, schedule and organization set in place.

*M. Rycroft (ed.), The Space Transportation Market: Evolution or Revolution?, 127–135.*
© 2000 *Kluwer Academic Publishers. Printed in the Netherlands.*

## 2.     Technology Evolution of H-IIA from H-II

The H-IIA program had been planned to evolve from  the  H-II ELV. The principal objectives taken from the H-II experience were to:

- Be robust in manufacturing and operations

- Improve reliability

- Prepare wider ranges of launch capability and mission flexibility

- Reduce recurrent costs by a factor of two or more.

To meet these requirements, major changes in the H-IIA design were:

**Augmented versions**

- A liquid rocket booster (LRB) of the same size as the core stage but with two LE-7A engines attached to the standard type to give 7.5 tonnes of GTO capability

- Another LRB attached could give a potential growth to 9.5 tonnes of GTO capability although it is not within the scope of the current H-IIA program

- The standard type has two enhanced versions, for 4.5 and 5 tonnes of GTO capability, with two and four small strap-on solid boosters (SSBs), respectively.

**First stage engine, LE-7A**

- Extreme reduction of welding, by utilizing forged materials, cast iron and bolt/flange joints

- Automatic welding rather than skilled hand welding

- Reducing the number of parts in the injector elements, coolant tubes of upper nozzle skirt, and so on

- Reducing the operating pressure of the main combustion chamber and operating temperature of the pre-burner and turbines

- Adding thrust throttling capability with a removable lower nozzle skirt of metal with cooling inside.

### Second stage engine, LE-5B

- Engine cycle change to a chamber expander bleed cycle from a regenerative nozzle expander bleed cycle, which enables engine tests at sea level to be carried out since the engine can operate without a nozzle skirt

- Combustion chamber structure changed to monolithic Cu alloy channel structure instead of Ni alloy coolant tube brazed structure.

### Propellant tank structure

- Monolithic oval domes manufactured by a plate spin-forming process used for first and second stage propellant tank ends instead of previous hemispherical domes assembled from panels welded together.

### Inter stage structure

- Monocoque sandwich structure of carbon fiber reinforced plastic (CFRP) skin and foam core used instead of previous skin-stringer structure of Al alloy.

### Solid rocket boosters

- Monolithic CFRP motor case of higher operating pressure used instead of previous four segmented steel cases, with each case assembled from panels welded together

- Electro-mechanical actuators for nozzle gimbaling used instead of hydraulic ones.

### Avionics

- Reduction of the number of parts by introducing highly integrated circuit parts, device integration and making the greatest possible use of CPU/software

- Introducing redundancies into such devices as CPUs, gyros and accelerometers.

**Ground support facilities**

- Reduction of launch operation work days, by a factor of three or more, by streamlining operational work share at the factory, launch site assembly building and launch pad, along with a new arrangement of the launch facilities.

## 3.    H-II Flight Anomalies and Investigations

### 3.1   F5/LE-5A Anomaly

H-II F5 lifted off from the Tanegashima Space Center on February 21, 1998 to inject the COMETS satellite into GTO. The flight was nominal up to 40 s into the second burn of the second stage. An anomaly occurred 1450 s into the flight; then, after seven seconds, the second engine, LE-5A shut down 140 s earlier than expected. As a result, COMETS was injected into a rather lower orbit than planned and the launch was not successful.

By analyzing the flight telemetry data, the NASDA investigation team estimated that the premature shut down of the LE-5A occurred as combustion gases leaked from the side of the lower part of the combustion chamber; then, the engine control power line was burned, which led to shut off of the main valves of the engine.

It was considered that the brazing between the cooling tubes in the lower part of the combustion chamber cracked, causing combustion gas leakage. Then the leaking hot gas heated up some of the cooling tubes and deformed them so that the crack widened. Contributory factors could have been a weak part due to uneven brazing and the existence of latent damage or deformation made during an unexpected low pressure firing at the acceptance tests before the flight, which was not detected by a special investigation after the firing. Both possible causes relate to specific requirements for better manufacturing quality management and inspections.

Although another LE-5A flight (that was H-II F7) was cancelled recently in relation to the F8 anomaly, some necessary corrective actions learned are to:

- Confirm more detailed brazing conditions by adopting advanced technologies such as a micro-focus X-ray inspection device with digital image processor

- Enrich the database of brazing materials and clarify the design requirements for brazing in order to establish a robust manufacturing process.

These actions are applied to the brazing process between the coolant tubes for the nozzle skirt of the LE-7A and LE-5B engines.

## 3.2    F8/LE-7 Anomaly

H-II F8 lifted off from the Tanegashima Space Center on November 15, 1999 at the scheduled time 16:29 JST to inject the MTSAT satellite into GTO. Significant flight events were as follows:

- The flight was nominal up to two thirds of the duration of the first stage boost phase

- The anomaly occurred 239 s after lift-off

- The LE-7 engine operating pressures in the main combustion chamber, pre-burner and fuel turbo-pump (FTP) outlet dropped abruptly to zero in about 0.5 s; axial accelerations also dropped to almost zero

- Some white gaseous jet was observed around the engine on ground videos, and the vehicle began to tumble

- Despite the main engine anomaly, vehicle sequencing continued through the first and second stage separation and ignition of the second engine (LE-5B) following the onboard computer commands

- The LE-5B engine burned normally despite the vehicle tumbling, for more than 100 s until the loss of telemetry data

- Destruct commands were sent to the vehicle at 459 s into the flight by a range safety officer; the impact area of the first stage and the second stage with the satellite was in the Pacific Ocean about 800 km downrange from the launch site, and about 380 km North West of the nearest island, the Ogasawara Islands of Japan.

An investigation team was established in NASDA; for five months collaborative investigations with a specialist group formed by the Space Activities Commission of the government, and also industry teams, took place.

Flight telemetry data were analyzed, and it was found that the FTP lost pressurization capability without any reduction of rotation speed early in the engine anomaly. This phenomenon, called "pump stall", caused a rapid and marked decrease of the hydrogen supply to the re-generative cooling path, pre-burner and main injector of the engine, and then led to an abrupt engine power down.

Through a fault tree analysis (FTA) method, four out of thirty possible component level causes supported by the telemetry data were found to be:

• Leakage or structural failure of hydrogen line downstream of the FTP

• Blockage of hydrogen dome of the main injector

• Blockage or leakage of hydrogen line upstream of the FTP

• Structural failure of FTP components.

After two months of survey effort in the Pacific Ocean, most of the components of the failed LE-7 engine were found on the seabed at about 3,000 m depth. And most of them were salvaged. It was discovered that the FTP inducer blades were heavily damaged, deformed and broken, and one of three blades was torn off. Many dense striations (trace of crack development) were observed on that broken surface by electron microscope photographs (see Fig. 1). It was found out that an FTP inducer blade was fractured by fatigue.

In order to evaluate the major factors which generate fluctuation stresses at an inducer blade, two types of tests were recently executed. One involved water flow tests using the same type of inducer as installed in LE-7 development models. Detailed pressure fluctuations in the inlet duct and inducer casing, oscillating strains on the inducer blades, and cavitation in the flow were measured and observed visually. The other was engine hot firing tests using a LE-7 flight model. Detailed pressure fluctuations and FTP rotor vibrations were measured by simulating hydrogen tank depressurizing control during the flight.

The factors responsible for the fluctuation stress at an inducer blade were considered to be:

• Oscillation forced by cavitation generated around the inducer

• Resonance of an inducer blade with a pressure oscillation in the liquid hydrogen inflow at a natural frequency of the blade.

**Figure 1.** Broken surface of FTP inducer blade— a scanning electron microscope photograph

It was learned that a cavitation phenomenon and related swirling back-flow with vortices around an inducer were rather more violent than considered previously, and that inducer cavitations were rather sensitive to its clearance and formation, especially at the blade leading edge.

A microscopic machining tool mark, which was allowed by the design standard, was observed at the crack initiation point on the broken surface; this could have promoted a concentration of stress there.

Taking account of new knowledge on cavitation phenomena, it is estimated that the inducer fatigue fracture on the H-II F8 was caused by fluctuating stresses over the fatigue limit, which were generated in an inducer blade under compounded worse conditions peculiar to that particular engine. Due to unbalanced rotation, the rotating inducer came into contact with the casing and led to FTP "stall"; then the engine stopped abruptly.

The inducer casing was broken up by a compound load of mechanical contact of inducer blades, inner pressure growth and thermal stresses, causing a massive discharge of liquid hydrogen from the open end of an inlet duct.

## 4.    Corrective Actions for the H-IIA Program

Responding to this painful experience, NASDA came out with significant corrective actions to the current H-IIA program.

In order to concentrate all efforts on the H-IIA program under development, the H-II program was canceled, leaving a flight, F7. In a more elaborate development, supplemental critical development tests, a flight test and schedule extension are incorporated. And in order to enhance the maturity of ELV technology, appropriate experimental data have to be acquired even after the development phase.

Detailed corrective actions are as follows:

• **Reinforcement of verification tests toward the first flight**

In order to validate marginal operation conditions as far as possible, supplemental tests of the LE-7A, LE-5B, SRB-A and other critical components are incorporated into the current development plan.

• **Reinforcement of flight verifications**

Another flight test is added; two test flights will be flown for verification and detailed data acquisition.

• **Unification of LE-7A configuration for the standard type of H-IIA**

As the current LE-7A has an issue of excessive side load at its starting and stopping transition when installed with a lower nozzle skirt of a metal sheet, it used to be planned that a complete nozzle skirt (for full expansion of the combustion gas flow) would be fitted after the first flight to which a "short skirt" (without the lower part of the nozzle skirt) was to be used. Then, in order to keep more time for validation of the counter measure to cope with the side load issue, which has been under testing for its robustness, and to avoid mixed developments of two LE-7A configurations, a "short nozzle" configuration would be commonly used for all the standard type H-IIA's, despite the performance being reduced by a few percent. An augmented type test flight, to which a full configuration nozzle skirt will be used, is scheduled for 2003.

• **Anti-cavitation measures, derived from the F8 anomaly**

The LE-7A engine has been designed to suppress cavitation and swirling backflow at the inlet of the FTP inducer, generated by depressurizing control of a hydrogen tank for the H-II/LE-7 case. Since the H-IIA/LE-7A does not have any flow rectifying vanes at the inlet duct, neither interference nor resonance between the hydrogen inflow, vanes and FTP inducer blades are expected. In order to validate the effectiveness of the design, specific inducer unit tests, FTP unit tests and LE-7A hot firing tests are being conducted.

• **Reinforcement of quality assurance activities**

In order to enhance the maturity of ELV technology, quality improvement programs are to be promoted even in the recurrent phase. These are:

- Technology validation tests for critical components of engines

- Stepping up the follow-on design and basic technology work

- Clarification of the detailed requirements for special manufacturing processes such as welding and brazing  peculiar to space launch vehicles.

# The Aerospace Vehicle of the Future

**G.E. Mueller, D. Kohrs,** Kistler Aerospace Corporation, 3760 Carillon Point, Kirkland, Washington 98033, USA

e-mail: gmueller@kistleraerospace.com, dkohrs@kistleraerospace.com

**Abstract**
Although the first successful orbital flight occurred over 40 years ago, the high-price of launch remains a significant barrier to the development of all space activities. The manufacturing costs of large, complex systems will always remain high, setting a lower limit on the affordability of expendable launch vehicles. Only a fully-reusable transportation system can dramatically reduce the cost of access to space. Although the initial cost of such a system is greater than for expendable vehicles of equivalent performance, the additional development budget can be recouped after only a small number of flights. The vehicle's reusability also drives reliability. Additional costs for more reliable systems are justified by flying the vehicle many times.
Kistler Aerospace Corporation was formed to realize the promise of reusability. Kistler is developing the K-1, the world's first fully-reusable aerospace vehicle, designed for up to 100 flights. The two stages of the K-1 both return to their launch site using parachutes and airbags. Each vehicle requires only minimal processing and refurbishment between successive launches. Kistler will maintain a fleet of K-1 vehicles capable of handling a high launch rate for routine access to space. The K-1 program is very different from other existing reusable launch vehicle development programs. Most importantly, the K-1 is not a technology development program. The K-1 relies solely on existing technologies proven in other successful aerospace programs. This reduces development cost, development time, and program risk. Kistler is also an entirely commercial venture with no government funding. This has led Kistler to develop a market-driven service at an affordable price. Kistler has raised over $ 500 million for the K-1 program, and plans to commence a full-flight test program in the near future from its dedicated launch site in Woomera, Australia. The K-1 is a medium-lift launch vehicle, designed to service a wide range of space missions in low Earth orbit and beyond. Kistler expects that the K-1's low price, flexibility and reliability will quickly enable it to capture a dominant market share in the medium and small-lift market once it enters commercial service. Initially, expendable launch vehicles will still be required to launch large payloads beyond the capability of the K-1. However, Kistler expects that the dramatic cost-savings offered by the K-1 will drive satellite operators and manufacturers towards smaller payloads. The aerospace vehicle of the future will clearly be reusable, and Kistler expects that the K-1 will be at the forefront of this revolution.

## 1.    Introduction

The cost of delivery to orbit is often greater than the cost of actually manufacturing and operating a satellite. Despite the current high price of orbital launch services, demand for end-use satellite services remains strong and many satellite ventures are quite profitable. If the cost of launch services were to decrease significantly, satellite services would become more profitable and more competitive with terrestrial communications alternatives, spurring even further demand for launches.

*M. Rycroft (ed.), The Space Transportation Market: Evolution or Revolution?*, 137–144.
© 2000 *Kluwer Academic Publishers. Printed in the Netherlands.*

Since the first commercial satellites were launched, however, the cost of delivery to orbit has not significantly decreased. The manufacturing costs of large, complex systems sets a lower limit to the affordability of one-shot, expendable launch vehicles. To stay competitive, expendable launch vehicle providers are adapting larger and larger designs to improve stage mass fractions, thereby reducing launch price on a per kilogram basis.

Despite these efforts, only a fully reusable commercial space transportation system, amortizing development costs over many flights, can significantly reduce the cost of access to space. A number of efforts are underway to develop such a system. Some of these efforts depend on the development of new, cutting edge technologies, and therefore have long development schedules, high risk, and high investment costs. One program, being developed by the Kistler Aerospace Corporation, takes a different approach, integrating proven technologies into a package designed to service a wide range of space missions.

## 2.    Overview of Kistler Aerospace Corporation

Kistler Aerospace Corporation was formed in 1993 by Walter Kistler, the co-founder of Kistler-Morse Corporation, and Bob Citron, the founder of SPACEHAB, Inc. Since 1995, Kistler has been led by Chairman Robert Wang and Chief Executive Officer Dr. George E. Mueller, the former head of NASA's Apollo Manned Space Program. Kistler Aerospace's mission is to develop and operate the world's first fully-reusable aerospace vehicles, called the K-1, designed to service a wide range of space missions. Kistler has assembled a team of preeminent aerospace experts and space program managers to design the K-1 vehicle and to manage the K-1 program. The design team members have collectively guided most of America's major space programs, including Redstone, Mercury, Gemini, Saturn, Skylab, Apollo, the Space Shuttle and the International Space Station. These experts also have extensive commercial space program experience. Kistler is fabricating a fleet of fully reusable K-1 vehicles that will provide customers with reliable and flexible launch capabilities at a competitive price. Kistler is an entirely commercial program with no government funding.

Kistler is leading the K-1 systems engineering and integration through an integrated team composed of Kistler and contractor personnel. Each of Kistler Aerospace's contractors is a leader in its respective field of the aerospace industry and has significant experience in the construction of similar components. This team consists of Lockheed Martin Michoud Space Systems, Northrop Grumman Corporation, GenCorp Aerojet, Draper Laboratory, Honeywell, Irvin Aerospace, Inc, and Oceaneering Thermal Systems.

The K-1 combines existing and tested propulsion technology with advanced lightweight composite structures to create a low-cost and reliable vehicle capable of delivering a variety of payloads to a wide range of altitudes and inclinations in LEO. Kistler plans to build a fleet of five K-1 vehicles which will provide a large annual flight capacity. Kistler believes that its use of existing technology and reusable modular components will enable the K-1 to achieve an operating cost not attainable by current expendable launch vehicle providers.

## 3.    The K-1 Vehicle

The K-1 reusable aerospace vehicle, shown in Fig. 1, has two stages. The overall vehicle is 36.9 m (121 ft) long and weighs 382,300 kg (841,000 lbs) at liftoff. The first stage, or Launch Assist Platform (LAP) is 18.3 m (60 ft) long, 6.7 m (22 ft) in diameter, and weighs 250,500 kg (551,000 lbs) at launch. The LAP has three main engines utilizing LOX and kerosene propellants. The second stage, or Orbital Vehicle (OV) is 18.6 m (61 ft) long, has a diameter of 4.3 m (14 ft), and weighs 131,800 kg (290,000 lbs) fully-fueled. Each stage carries its own suite of redundant avionics and operates autonomously.

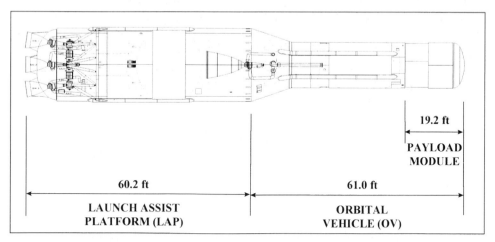

**Figure 1.** Profile of the K-1 vehicle

Fig. 2 shows a flight profile for the K-1 vehicle. Following stage separation at approximately 43 km altitude, the LAP reorients itself and restarts its center engine to return the stage to the launch site. The LAP lands near the launch site using parachutes and airbags.

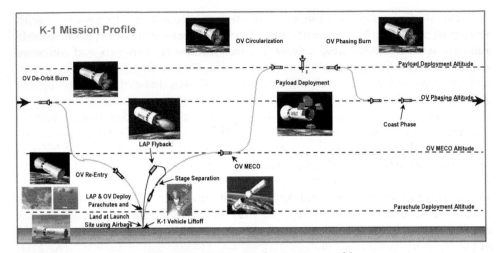

**Figure 2.** K-1 typical mission profile

The second stage, called the Orbital Vehicle (OV) ignites a single engine utilizing LOX and kerosene propellant following stage separation. The OV uses an Orbital Maneuvering System (OMS) engine for circularization, phasing, deorbit, and collision avoidance burns. The Payload Module attached to the top of the Orbital Vehicle contains the payload and dispenser. Every part of the Payload Module and dispenser returns to Earth intact. After deploying its payload, the OV coasts for approximately 22 hours before returning to the launch site using parachutes and airbags.

Fig. 3 shows the performance of the K-1 vehicle from its Spaceport Woomera launch site.

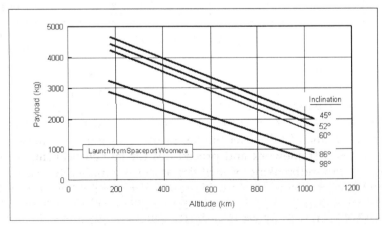

**Figure 3.** Circular orbit performance

More information on the K-1 vehicle, including interfaces and payload environments, can be found in the *K-1 Payload User's Guide*, [Reference 1]. Launch environments and interfaces are similar to those found in most expendable launch vehicles.

## 4.    Launch Sites

Kistler has selected two sites from which it will launch and land its K-1 vehicles: (i) the Spaceport Woomera site in South Australia, and (ii) the Nevada Test Site. The locations and initial launch corridors of each site are shown in Fig. 4. Kistler selected Spaceport Woomera and Nevada because the K-1 flight profile dictates launch and landing over land and this is achievable most safely from a remote area with low population density and with the necessary available infrastructure.

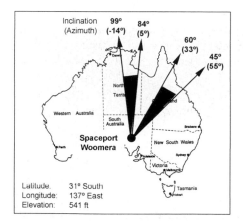 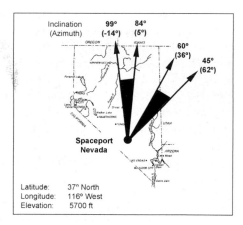

**Figure 4.** K-1 launch sites

On August 28, 1998, the Australian Department of Industry, Science and Tourism and Department of Defence executed the Operations Agreement authorizing Kistler to use the Woomera Prohibited Area for construction and operation of a launch facility, set the terms and conditions for those activities, and established a framework and procedures for launch licensing and launch operations. A groundbreaking of facilities at the Australia Site was on July 23, 1998. The Spaceport Woomera site design is complete and site permits have been obtained. Kistler plans to begin construction of the Nevada Site after successful demonstration of K-1 flight operations in Australia.

## 5.     Progress to Date

Kistler has made substantial progress toward the development of the K-1 vehicle and expects to commence test flights in the near future from Woomera, Australia. To date, 75% of the vehicle is complete by weight. The First and Second stage LOX tanks are completed and the NK-33 certification engine has completed six successful tests (see Fig. 5). At present, 46 NK-33/NK-43 engines are available at Aerojet. There are 37 NK-33 and 9 NK-43 which support 180 missions. These engines have the highest thrust to weight ratio of any existing engine in the world. Our main parachute development program is complete and 21 out of 23 main structure panels are completed. The majority of our avionics hardware has been delivered, Guidance, Navigation and Control (GN&C) software is complete and hardware-in-the-loop testing at Draper Labs is ongoing. The Launch Facility design is 100 % released and the launch site construction contract has been awarded.  Kistler has raised over $ 500 million in private equity and has insured its flight test program for the replacement value of the K-1 vehicle through a global consortium of underwriters.

**Figure 5.** LAP LOX tank roll-out and engine test firing

## 6.     The K-1's Competitive Edge

### 6.1   Kistler's Design Approach

The K-1 has a number of important advantages over all expendable launch vehicles and reusable launch vehicles in development. These advantages include low-cost, enhanced reliability, and flexible scheduling. They arise not only from the K-1's reusability, but also from Kistler's design philosophy. The K-1 makes use of mostly proven technologies and off-the-shelf hardware.

## 6.2   Low Cost

The most obvious advantage of reusability is dramatically lower cost. Kistler has published a list price of $ 17 million for a basic K-1 LEO launch service. Although the K-1 is a medium-lift launch vehicle, this list price is competitive with the launch prices offered by vehicles in the small-lift class. For customers desiring delivery beyond LEO (such as GTO and GEO), Kistler will offer the option of an expendable Active Dispenser.

## 6.3   Enhanced Reliability

Reusability enhances reliability. In order to maximize the return on investment, reusable vehicles are designed with more reliable and robust systems. For example, the K-1 is equipped with a triplex avionics architecture and its structure is designed to higher factors of safety than commonly used in the design of expendable launch vehicles. The additional cost for these systems is more than offset by using them many times. The K-1 is designed for reuse up to 100 times, with a minimum of refurbishment and component replacement over a vehicle lifetime. Reuse of the same flight proven hardware improves reliability.

## 6.4   Flexible Scheduling

Kistler offers its customers considerable flexibility in scheduling their launches. Kistler will build a fleet of K-1 vehicles and maintain several at each launch site. Each vehicle will be processed through successive launches in as little as nine days. Kistler will therefore have a tremendous flight rate capacity not dictated by a backlogged manufacturing schedule, as is the case with expendable launch vehicles. Kistler also operates its own dedicated launch sites and processing facilities instead of sharing facilities with other launch vehicle providers. With a fleet of vehicles operating at full capacity, Kistler can launch many times a year from each launch site.

For successive flights of the same spacecraft type, such as satellite constellations, Kistler can offer true launch on demand capability. Satellite customers will be able to keep spares on the ground, instead of aloft, and call them up for a K-1 launch only when needed. With spares in storage at the launch site, the K-1 can complete delivery to orbit in 30 days or less for a replenishment launch.

## 7.    Conclusion

Only a fully reusable space transportation system can reduce the cost of access to space enough to spur on an explosive growth in space enterprise. Kistler Aerospace Corporation's K-1 vehicle will be the first such system. Kistler has combined proven technologies into a package that can service a wide range of space missions. Significant progress has been made in hardware development and component testing. Kistler expects to commence a flight test program in the near future from its dedicated launch site in Woomera, Australia, and begin commercial operations shortly thereafter.

The K-1 has tremendous advantages over expendable launch vehicles in terms of cost, reliability, and schedule flexibility. Kistler expects that, with this competitive edge, the K-1 will quickly capture a significant market share and become the market leader shortly after commercial operations commence.

**References**
1.    Kistler    Aerospace    Corporation:    *K-1  Vehicle  Payload  User's  Guide*, <http://www.kistleraerospace.com/payload/payload.htm>.    Debra    Factktor Lepore, May 1999

# The Delta Launch Services Mission Integration Process — Improved, Cost-Effective Access to Space

T. Morrison, D. Festa, The Boeing Company, Expendable Launch Systems, 5301 Bolsa Ave., Huntington Beach, California 92647, USA

**Abstract**
Utilizing launch vehicle technology is a complex undertaking, often requiring years of preparation before a launch. Mission integration, the process of linking a payload to the launch vehicle, typically requires up to two years of preparatory work. This work must be coordinated between the launch service provider, the payload manufacturer, and the owner/operator of the payload. Improving the mission integration cycle could decrease the time required to prepare for a launch, and the overall cost associated with the mission.
    This paper examines the processes required for mission and launch site integration using the Boeing Delta family of launch services. The mission integration process demonstrates how improvements in technology and lessons learned have been incorporated into the Delta IV family. Finally, additional areas of improvement are suggested for future launch vehicles in an effort to promote low-cost access to space through improvements in the mission integration process.

## 1. Delta Overview

### 1.1 Delta Launch Services and the Delta Family of Vehicles

Delta Launch Services, Inc. (DLS), a wholly owned subsidiary of The Boeing Company, provides safe, accurate, and affordable deployment solutions for satellite owners, operators, and manufacturers. Building on forty years of heritage and experience, the DLS fleet currently is composed of the Delta II, Delta III, and Delta IV launch systems (Fig. 1). With more than 275 missions flown to date, Boeing mission integration draws upon a wealth of experience, knowledge, and capability that is among the best in the world. Boeing has experience integrating commercial, government, and scientific payloads that span the entire range of mission requirements, including low-Earth orbit, geostationary missions, and interplanetary exploration. Boeing uses launch pads at Cape Canaveral Air Force Station (CCAFS), Florida, and Vandenberg Air Force Base (VAFB), California, for all Delta missions.

### 1.2 Delta II — The Workhorse of the Constellation Industry

Delta II is one of the world's most reliable and respected launch systems. With its sound design, high reliability, and pinpoint orbital accuracy, Delta II provides an excellent choice for successful launch and deployment of small- and medium-class payloads. The Delta II launch vehicle was introduced into service in 1989, and has since recorded 90 missions with a 97.8 % mission success rate.

145

*M. Rycroft (ed.), The Space Transportation Market: Evolution or Revolution?*, 145–155.
© 2000 *Kluwer Academic Publishers. Printed in the Netherlands.*

HB00943REP0.2

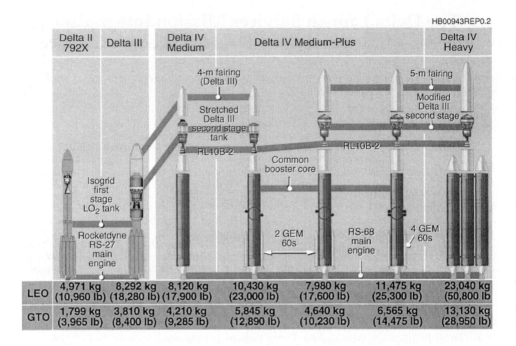

**Figure 1.** The Delta family of launch vehicles

The Delta II vehicle can be configured to meet specific mission needs. Customers may choose the type of payload fairing in which to encapsulate their payload, the number of strap-on solid rocket motors required for optimized performance, and an optional third stage. Delta II has been the workhorse of the constellation industry, launching 55 satellites for Iridium® (5 satellites per launch) and 28 satellites for the Globalstar™ constellations (4 satellites per launch), with 100% success. Delta II also has launched 31 satellites for the Global Positioning System (GPS). Boeing continues to evolve the Delta II by introducing additional capabilities over the next two years. The inaugural dual-manifest mission and a lengthened version of the 3.05-m (10-ft)-diameter composite fairing will be launched in 2000. The new Delta II Heavy, using more powerful solid rocket motors, will be introduced into service in 2001.

### 1.3 Delta III – Extended Capability to Meet Customer Needs

Delta III has evolved from Delta II to increase significantly the lift capability needed to support today's launch services market. The Delta III launch vehicle combines proven Delta II elements with new, flight-proven components to ensure mission success. Delta III uses a modified Delta II first stage and avionics, while featuring improved solid rocket motors for thrust augmentation. In a change from Delta II, a cryogenic second stage is added for increased performance. The second stage is powered by a single RL10B-2 Pratt & Whitney liquid-hydrogen/liquid-oxygen engine derived from the highly reliable RL10 engine family. The 4-m-diameter composite fairing provides an increased payload envelope over Delta II's 3.05-m-diameter composite fairing.

### 1.4   Delta IV – An Evolved Family of Expendable Launch Systems

Delta IV continues the evolution of Boeing launch systems. Designed in response to US Government and commercial specifications, the Delta IV family of vehicles is anticipated to reduce launch costs by 25 % or more, while maintaining or improving reliability. There are two primary variations of the Delta IV vehicle — the Delta IV Medium, and the Delta IV Heavy. These two variations were selected by the U.S. Air Force's Evolved Expendable Launch Vehicle (EELV) program to be the majority launch service provider (19 out of 28 possible launches) for future U.S. government launch needs. In addition, Boeing is developing concurrently three commercial derivatives of the Delta IV Medium, known collectively as the Medium-Plus series. Delta IV is progressing toward a first launch in 2001.

All variations of Delta IV are based on a common booster core (CBC) first stage, which allows for a low-cost and simple design. The CBC is powered by the new RS-68 cryogenic main engine. The second stage is derived from the Delta III second stage, with increased propellant tank capacity. A composite fairing (either 4 m or 5 m in diameter) encapsulates the payload and protects it during initial phases of flight. The Medium-Plus series will also feature two or four solid rocket motors for thrust augmentation.

The Delta IV Heavy vehicle uses a central CBC as the first stage, with two additional CBCs attached for thrust augmentation. The second stage and payload fairing are both 5 m in diameter, and capable of launching two large satellites simultaneously, in a dual-manifest configuration.

## 2.   Mission Integration

### 2.1   Overview

The mission integration process developed by Boeing has evolved over the past forty years in working with customers (both commercial and government) to integrate payloads with Delta launch systems. This process is designed to support the requirements of the payload as well as the launch vehicle. We work closely with our customers to tailor the integration activity to meet individual requirements.

The integration process begins when Boeing receives authority to proceed (ATP) from the customer, extends through payload construction, processing, and launch, and concludes when mission success has been determined. The customer can be either the payload manufacturer or the eventual satellite owner/operator, or both. The relationship between the manufacturer and the owner/operator must be established from the outset, but Boeing works with all parties to ensure total mission success.

## 2.2   Mission Integration Schedule

The mission integration schedule varies from launch to launch based on mission-specific needs. Boeing has established a mission integration process for use with all Delta launch vehicles (Fig. 2), with a step-by-step progression of activities leading up to launch. This baseline mission integration process ranges from 104 weeks to 91 weeks long. Mission-unique features and launch schedule requirements must be evaluated to establish the final mission integration timeline. For example, Boeing was able to accelerate the mission integration process for three repeat Globalstar launches, and successfully integrated and launched 12 satellites within a one-year period.

As time-to-market becomes increasingly important to our customers, Boeing strives to reduce the length of time of the mission integration process. At the same time, mission success is never compromised. These timelines depend on a mutual effort by Boeing and the customer to provide the necessary inputs into the process at the appropriate times. If a customer requests a more compressed schedule, the customer must provide Boeing all the inputs necessary to support the accelerated integration process.

## 2.3   Managing the Integration Process

To manage the integration process, Boeing assigns a mission integration manager (MIM) to provide the customer dedicated support and overall mission integration accountability. The Boeing MIM is responsible for establishing and leading the Boeing mission integration team. At the same time, the customer also assigns a spacecraft MIM to coordinate all customer activities with the Boeing MIM. These two managers are the primary focal points for all mission integration activities.

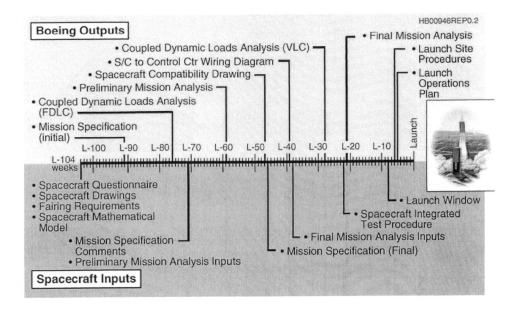

**Figure 2.** Sample mission integration process

The Boeing mission integration process is based on a team approach (Fig. 3). Key team members include the customer, Boeing MIM, Boeing integration engineer, Boeing launch site manager, and Boeing program manager. The integration engineer is responsible to communicate mission requirements to the analysis, hardware, and test teams. The launch site manager implements necessary mission requirements at the launch site, including payload processing facility, range coordination, and launch pad activities. The program manager provides business management support, while monitoring overall progress according to the launch services contract.

A major task for the MIMs is to ensure that all lines of communication function effectively. To this end, meetings are held throughout the course of the integration process to ensure that all parties are kept up-to-date. This includes periodic technical interchange meetings, working group meetings, and formal readiness reviews.

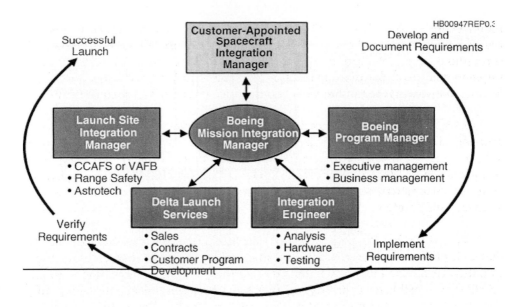

**Figure 3.** The mission integration manager oversees all Boeing aspects of the integration process

## 2.4  Boeing and Customer Responsibilities

The mission integration process is a joint effort between Boeing and the customer, and requires a coordinated effort by all parties to ensure mission success. Fig. 2 shows the general responsibilities of each party in developing a successful mission integration schedule.

The entire process begins with a mission integration kickoff meeting, where Boeing and customer program personnel are introduced and preliminary mission requirements are set. At this point, the customer completes a detailed spacecraft questionnaire (a copy of which can be found in the Delta Payload Planners Guides [see Reference 1]). This questionnaire documents initial payload characteristics and requirements, and is used to develop the first draft of the interface control document (ICD), also known as the mission specification.

The ICD has evolved into a master tool to capture all mission-specific requirements necessary to execute mission integration objectives successfully. The ICD contains the payload description, launch vehicle induced environments, payload compatibility drawings, targeting criteria, special requirements, etc.. This ICD is used to communicate mission-specific requirements to the Boeing mission integration team, and is updated throughout the integration cycle to ensure that all requirements are implemented. A verification matrix is developed to identify what methods (e.g., testing, analyses) were used to verify compliance with the ICD requirements.

Once the mission specification has been completed, Boeing performs a number of analyses to verify compatibility of the payload with the launch vehicle, and to ensure proper production of launch vehicle hardware and configuration of ground support equipment. These include coupled dynamic load analyses, fairing dynamic envelope verification, preliminary/final mission analyses, and safety packages, as well as the launch operations plan.

Once at the launch site, Boeing works closely with the customer to ensure that activities at the payload processing facility and the launch pad flow smoothly. A Boeing launch site manager, working in close coordination with the MIM, is responsible for coordinating all activity between the launch vehicle and the spacecraft, including mating to the payload attach fitting (PAF), transporting the payload to the launch pad, encapsulating the payload, testing, and checkout.

As a final product of our safety analyses, the launch vehicle missile system prelaunch safety package (MSPSP) is combined with the spacecraft MSPSP (provided by the customer) and delivered to the range safety office at the launch site for approval. When all approvals have been received, the US Department of Transportation issues Boeing a launch license, and all is ready for liftoff.

## 2.5  Mission Success

The mission integration effort does not end with the launch of the payload. Boeing continues to work with the customer after launch to verify that all mission requirements were met. During flight, a comprehensive instrumentation package collects flight data for all critical components and systems. These instruments monitor vehicle performance and environments that the payload has been exposed to, from liftoff to payload separation. These assessments are shared with our customers to provide an objective review of vehicle performance data during the mission.

Boeing focuses on mission success by meeting or exceeding customer expectations. Mission success is determined by three criteria. First is the placement of the payload into the required orbit. Second is that telemetry data show that the mission requirements were met. And finally, telemetry and satellite data are analyzed to confirm that the launch vehicle had no adverse effects on the payload during deployment or thereafter.

## 3.    Launch Operations

Over the past forty years, Boeing has continuously evolved its efforts to streamline and simplify the launch operation process. This evolution has resulted in significant savings of time and material for the benefit of the customer, while retaining high standards and flexibility to meet mission needs.

One example of our continuing evolution is the number of days before a launch that the payload is required for integration with the launch vehicle. In the 1970s, it took an average of 40 days of processing the payload on the launch pad for a mission to be ready. By the 1990s, Boeing had reduced that average to 24 days of on-pad processing and, with Delta IV, we are targeting 6 to 8 days (Fig. 4). These savings in turn mean that the payload can be made ready for launch in as little as 8 to 10 days before a launch, resulting in considerable savings to the customer. This is accomplished by encapsulating the payload at the payload processing facility, as opposed to doing so on the launch pad.

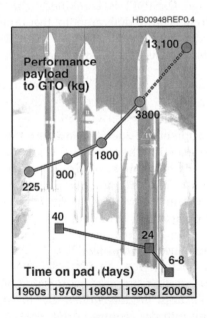

**Figure 4.** Delta's commitment to continuous improvement

Encapsulating the payload at an off-site location permits the launch vehicle to be integrated on the launch pad simultaneously, thus reducing the time required for the satellite to be at the launch pad. This also provides improved access to the satellite during processing, greatly facilitating the mission integration process.

Another example of streamlining the launch operation process is the horizontal integration of the Delta IV launch vehicle at an off-pad location (as opposed to being "stacked" on the launch pad). The CBC first stage is mated with the second stage at the Boeing horizontal integration facility (HIF) and then rolled out to the launch pad and erected. Once the vehicle is erected, the solid rocket motors (if required) are attached to the CBC. For the Delta IV Heavy, which requires three CBCs joined together, all three are integrated and joined together at the HIF and then transported as a unit to the launch pad. The encapsulated payload is transported from the payload processing facility, elevated to the top of the launch platform, and then attached to the upper stage (Fig. 5). When the entire vehicle has been integrated on the launch pad, fueling operations begin, and a countdown is initiated.

Horizontal integration of the Delta IV vehicle greatly reduces the operational time that is required on the launch pad. This is a significant improvement over traditional launch pad operations, as most overhead hoisting operations are eliminated. This increases personnel and equipment safety and reduces vehicle exposure to undesirable weather conditions.

## 4.    Delta IV Infrastructure Improvements

With the introduction of Delta IV, Boeing had the opportunity to reexamine all of its mission integration processes. Many of the processes used on Delta II and Delta III were retained, while others were improved or updated. For example, Boeing has instituted a centralized, interactive information system that permits personnel to manage data across the entire program. The GENISYS network allows designers, manufacturers, suppliers, and launch site operations personnel to share the latest information quickly. This greatly improves our ability to communicate design changes or updated information in all aspects of vehicle production and integration.

Another improvement comes from the Delta IV Focused Factory, located in Decatur, Alabama. This factory uses lean engineering principles to produce a large, complex system in one location. The focused factory is designed so that raw materials enter through one door, and a complete, fully validated subsystem rolls out through another door. Having all the equipment and facilities in one location allows for improved vehicle manufacturing times, which in turn helps support reduced integration cycles. The focused factory also manages the integrated assembly and checkout phase, where vehicle verification is performed.   This is done at the factory itself, where all the machinery and tooling are available to fix any anomalies that are found.

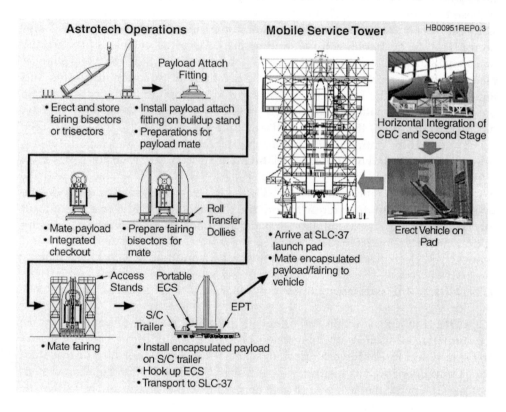

**Figure 5.** Delta IV payload encapsulation and launch operations

## 5.    Improvements for the Future

In an effort to improve our systems and processes, Boeing continues to optimize the mission integration timelines while maintaining mission success. Boeing is coordinating efforts with suppliers and partners to reduce lead times and become more responsive to market changes. For example, some items must be ordered from suppliers several months before they are needed. Boeing is working with all its partners to cut down on the number of long-lead time articles, and work more in parallel with other mission integration activities.

Boeing also is working to implement powerful computational tools that will decrease the labor-intensive analysis work that must be done before a launch. These tools will allow designers to run more iterations and to incorporate design changes more efficiently than ever before.

As the capabilities of Delta launch vehicles have increased, new satellite designs have been developed to meet the increasing demands of today's marketplace. Boeing is working with all the major satellite manufacturers to better understand each satellite "bus" to ensure full compatibility with Delta launch services.

## 6.   Summary

Continuously improving our mission integration process plays an important role in supporting the goals of low-cost access to space. By starting with a proven baseline, it is possible to minimize the impact of mission-unique requirements, while taking advantage of synergies associated with similar or repeat missions.

Mission integration is a key component of launch services that is crucial to mission success. With a proven track record of mission success, the Boeing mission integration team has repeatedly proven itself as thorough and capable of meeting mission requirements. The processes involved have been refined over the past forty years with a wide range of customers, including commercial, government, and scientific payloads. Boeing has adapted this experience to meet individual customer needs and is working with customers around the world to ensure mission success.

### References

1.   Specific details of the mission integration process, including schedules, available services, and roles/responsibilities of each party can be found in the Delta Payload Planners Guides. These guides provide detailed information on the vehicle performance capabilities and services provided by Boeing. These guides can be found on the web at <http://www.boeing.com/delta>.

# Reusable Launch Vehicles (RLVs) and Cost Engineering Principles

**D.E. Koelle,** TCS-TransCostSystems, Liebigweg 10, D-85521 Ottobrunn, Germany

e-mail: dekoelle@transcost.com

**Abstract**

The paradigm change in space launch vehicle design and operation - from performance optimization to minimum cost strategy - has greatly increased the importance of cost engineering. This is even more important for the future generation of reusable launch vehicles (RLVs) due to the relatively high development costs. The paper presents several examples of applied cost engineering - the selection of the cost-optimum vehicle concept, the reduction of development costs to some 60 % of previous business-as-usual costs, or even down to about 30 % by pure commercial-industrial project implementation; it also considers the conditions for achieving minimum space transportation costs with reusable launch vehicles. Finally, the role of cost engineering for the selection of technologies is described: it is important to verify the fact that the new technology will contribute to cost reduction - and not lead to increased costs and complexity.

## 1. Introduction

A major change has taken place in the area of space transportation in the past decade. Space launch vehicles are no longer a subject of national prestige and technology pride; they are now the subject of commercial operation and international price competition. This means that the basic paradigm for launch vehicle design and development has changed substantially, from the historical performance-optimized design (maximum payload by the most advanced technology with the lowest weight) to minimum-cost design and operations.

Unfortunately this fact has not yet been widely recognized. Especially in Europe there is little understanding or use of the rules and tools of cost engineering. Simply stated, cost has to be taken into account as a design and selection criterion already in the initial vehicle design phase. This is feasible nowadays with such tools as the Transcost-Model, for example. However, engineers are not (yet) educated this way; they still consider maximum performance and minimum weight as the primary goal. Although this is a great engineering challenge, it does lead to high costs.

What is required in today's commercial and competitive world of launch services is the minimum-cost approach. From the customer's (satellite operator's) viewpoint the launch vehicle type and design are secondary. What counts are transportation cost ($ per kg) and reliability. The application of cost engineering principles results in quite different launch vehicle design features,

157

*M. Rycroft (ed.), The Space Transportation Market: Evolution or Revolution?*, 157–166.
© 2000 *Kluwer Academic Publishers. Printed in the Netherlands.*

such as non-optimum staging, adaptation of existing engines (even with higher thrust levels than required) or higher weight, but lower development costs and lower costs per kg to orbit.

Early application of cost engineering principles are already apparent for expendable launch vehicles like the Japanese H2A (derived from the H-II only for reasons of launch cost reduction), as well as the Evolved Expendable Launch Vehicle concepts in the USA (ATLAS V and DELTA IV), with simpler stages and fewer engines and boosters. A more specific example in the rocket engine area is the RS-68 engine compared to the Space Shuttle Main Engine (SSME): It has a simpler design (fewer parts), lower pressure levels and greater thrust margin — leading to higher weight and lower performance — but also to lower costs and higher reliability.

The strict application of cost engineering principles for the design and definition of future Reusable Launch Vehicles (RLVs) is mandatory due to the fact that the development costs are relatively high. The Cost-per-Flight will finally determine the competitiveness of the vehicle and — in case of commercial financing — also the Return-on-Investment.

## 2.    RLV Concept Selection using Cost Engineering Principles

Numerous studies have been made of reusable launch vehicle concepts. However, the emphasis was placed on specific designs, new techniques and new technologies — but they have not been justified by cost advantages compared to alternative solutions. Based on advanced engineering ideas like air liquifaction in flight, or tripropellant rocket engines etc. (discussed in Section 4), extremely expensive concepts have even been proposed.

The discussion is still going on and has concentrated on the feasibility of SSTO (Single-Stage-to-Orbit) vs. TSTO (Two-Stage-to--Orbit) vehicles to LEO. However, there are tremendous differences between the various SSTO launch vehicle options with respect to technology and development costs:

(1)    Ballistic Rocket Configuration (VTOL) — with vertical take-off and landing, as tested by the Delta Clipper Demonstrator Vehicle,

(2)    Winged Rocket Configuration (VTO-HL) — with vertical take-off and horizontal landing, as the Space Shuttle Orbiter

(3)    Winged Rocket Configuration (HTO-HL) — with horizontal take-off, using a launch sled or a carrier aircraft, with the British HOTOL as an example, and

(4)  Winged Configuration with combined airbreathing/rocket propulsion, employing scramjet propulsion, with the US NASP Concept as an example

In addition, there are the most preferred TSTO Concepts:

(5)  Parallel-staged Winged TSTO Rocket Concept — sometimes with equal-sized stages, VTO-HL, and

(6)  Horizontally launched TSTO with airbreathing propulsion in the first stage and rocket propulsion for the second stage — with the German SÄNGER Concept as an example.

The concepts (2), (1), (5) and (6) are shown in Fig. 1 to about the same scale for a LEO paylad of 7000 kg.

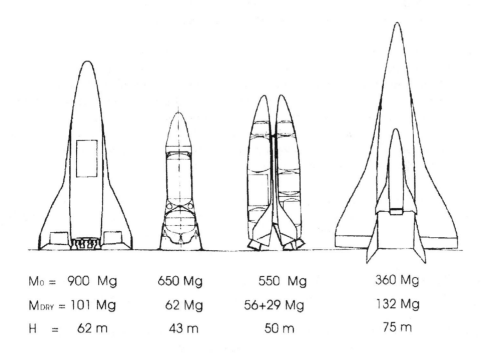

| $M_0$ = | 900 Mg | 650 Mg | 550 Mg | 360 Mg |
|---|---|---|---|---|
| $M_{DRY}$ = | 101 Mg | 62 Mg | 56+29 Mg | 132 Mg |
| H = | 62 m | 43 m | 50 m | 75 m |

**Figure 1.** Comparison of major RLV configurations with the same payload

The first three vehicles shown employ rocket propulsion only, the vehicle on the right uses airbreathing engines in the first stage. This SÄNGER-type vehicle provides the unique opportunity for take-off from Europe (at an airfield) as well as being the prototype for a Mach 4 passenger aircraft. Because of the hydrogen-fueled turboramjet engines the vehicle size is not smaller but larger than the rocket-propelled options, in spite of the low total mass at launch.

The development costs for the four vehicle concepts of Fig.1 as government-funded projects are as follows:

(1) Winged SSTO              12   billion  USD

(2) Ballistic SSTO            8   billion  USD

(3) Winged TSTO              16   billion  USD

(4) Winged TSTO/HTOL         23   billion  USD

In case of commercial development, these costs could be reduced to less than 40 % of the values shown, as discussed in Reference 1.

Even higher development costs than for version (4) must be expected for an airbreathing single-stage vehicle (NASP-type with scramjet propulsion) which is not shown here because its feasibility is doubtful and the economics are questionable.

The development costs presented here and the key cost drivers have been derived with the Transcost-Model (Reference 1) which allows the vehicle development, fabrication and operations costs to be determined without a detailed vehicle design at the subsystem level.

The great development cost differences of the launch vehicle options clearly show how important is the role of cost engineering in the selection of the vehicle concept to be developed in the future. For the more conventional TSTO vehicle, the cost will be some 30 % higher than for an SSTO vehicle (in spite of a 40 % lower launch mass). For a winged vehicle with the more conventional horizontal landing mode the development costs will be some 50 % higher than for a ballistic vehicle with vertical landing. Although this has been demonstrated successfully, it meets scepticism because it is not the usual mode of landing.

For most missions, especially to geostationary transfer orbit or for Earth-escape missions, two-stage vehicles will be required. Therefore, the question is not really SSTO vs. TSTO but whether the first stage should have orbital capability and can return to the launch site without additional fly-back provisions. Those features are required in case the first stage has suborbital velocity: in most cases a winged vehicle is proposed with airbreathing turbojet engines for the return flight to the launch site. This, however, results in a complex first stage design which has to fulfill the requirements of a VTO launch vehicle stage and those of a cargo aircraft. Experience has shown that the aircraft design requirements are mostly underestimated by launch vehicle designers, while the launch vehicle problems are normally under-estimated by aircraft designers. In any case it is a complex and expensive vehicle development. In addition there is the problem of stage separation and the increased effort for fly-back operations.

In principle, single-stage systems do have an inherently higher reliability vs. two different systems plus stage separation. From the cost and economics standpoint, therefore, an SSTO launch vehicle, i.e. a vehicle with an orbital first stage, is clearly the preferred choice.

## 3.    RLV Design using Cost Engineering Principles

The pre-requisites to arrive at a cost-efficient launch vehicle design are:

•    The recognition that cost is the premier criterion for future RLVs

•    Experience and technical judgement in launch vehicle design, and

•    The use of an analytical cost model tool.

The first trade-off to be made is the optimum vehicle size with respect to the required mission performance (payload). The conventional minimum dry mass and maximum engine performance goal definitely leads to the most expensive vehicle design. The application of cost engineering with a trade-off between weight, reliability and costs will result in a somewhat larger and heavier vehicle, but a vehicle with lower cost and higher reliability.

Besides the feasibility the performance sensitivity of an SSTO vehicle is being used as an argument against this most cost-efficient vehicle concept. However, there has not only been much progress in launch vehicle technology in recent decades, as explained carefully with many examples in Ivan Bekey's White Paper (Reference 2), but, in addition, the feasibility of a Reusable SSTO Vehicle is not just dependent on the application of advanced technology, but very much also on the vehicle size, as it is illustrated in Fig. 2.

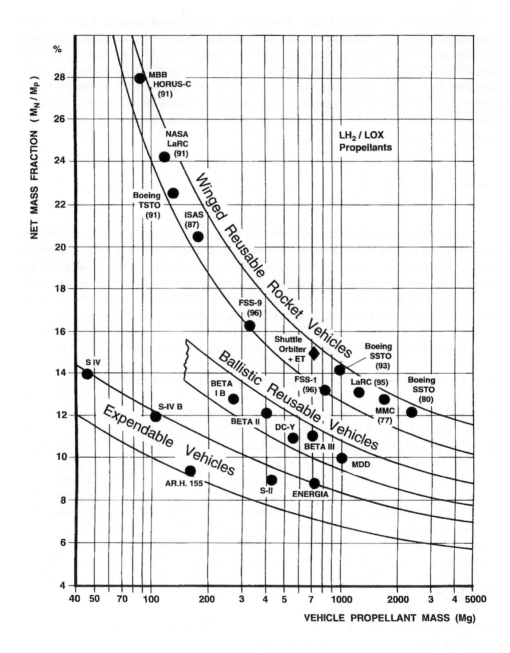

**Figure 2.** Net mass fraction (total mass minus ascent propellant mass and payload) decrease vs. launch vehicle size (ascent propellant mass)

All launch vehicle concepts show a distinct trend of decreasing Net Mass Fraction (NMF) for increasing size (e.g. propellant mass). This is more pronounced for reusable vehicles than for expendables, and it is most sensitive for winged vehicles. The NMF is defined as ratio or percentage of net mass vs. (ascent) propellant mass. The difference between dry mass and net mass is the propellant mass required for orbital operations (injection, braking impulse, attitude control) and the landing maneuver, as well as for residuals and as reserve propellants.

The decreasing NMF with increasing size means that the performance (payload) of vehicle is growing over-proportionally with size. This fact is often ignored in making comparisons or setting simple mass ratio requirements.

The basic result of this fundamental Net Mass Fraction trend vs. vehicle size is the cost engineering rule to overdesign the vehicle in order to achieve a comfortable performance margin at low cost. Fig. 3 shows the mass and cost sensitivity of SSTO Vehicles: for example, a 25 % increase of dry mass results in some 40 % higher payload, but only a 12.5 % higher development cost. Alternatively, increasing the launch mass (GLOW) by 10 % provides a payload gain (or margin) of some 18 % at only a 3 % higher development cost, assuming that the same technology is applied.

The other option is to keep the vehicle payload capability at a given value (including margin) and to increase the RLV dry mass and GLOW. In this case the development costs are not increased but reduced through application of more conventional and lower-cost technology and techniques. Thus, in general, a careful trade-off is required for the RLV design with respect to the optimum GLOW value, with a sufficient payload margin and minimum development costs.

Based on the NMF-trend vs. size it is also possible to derive the minimum realistic launch mass or GLOW (Gross Lift-Off Mass) for ballistic SSTO vehicles (Reference 3): For LEO (only) missions the minimum ballistic RLV mass is some 350 Mg, for the Space Station Orbit some 450 Mg, and for elliptical polar orbits some 600 Mg. For winged vehicles these GLOW values are about 50 % higher.

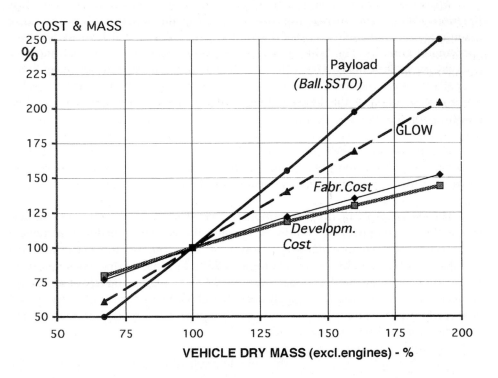

**Figure 3.**  RLV mass and cost sensitivities: large payload gain is feasible at small cost increase  (Ballistic SSTO vehicles)

## 4.    Technology and Cost Engineering

Cost engineering principles should also play a major role in technology assessment. So far, technology developments have been based only on the maxim of higher performance and lower weight. Many brilliant technical ideas have been produced, and (sometimes) millions of dollars invested in research and feasibility demonstrations before finding out that they do not improve the space transportation cost effectiveness, but only lead to higher complexity and higher costs.

One such idea was for a tri-propellant rocket system: using kerosene in addition to hydrogen as the fuel could reduce the required vehicle volume. Indeed the vehicle geometry can be reduced and the performance (payload) can be increased by some 3 %. However, the propellant system complexity and cost are increased by 50 % (!); the rocket engine design becomes more complex and more prone to failures. The same 3 % payload increase can be achieved easily, and almost without additional cost, by increasing the launch mass of the bi-propellant vehicle by 3 %.

Another example of advanced technology is the suggested use of air-breathing engines with air liquifaction during flight. During cruise flight in the lower atmosphere air is collected, liquified and separated to make liquid oxygen for the subsequent rocket propulsion phase. This means that liquid oxygen would not need to be carried on board at launch. However, complex and heavy equipment for liquifaction would be required onboard, with its inherent reliability being less than 100 %. Compared to this expensive machinery, liquid oxygen is cheap and most reliable.

*TCS-TransCostSystems*

## Cost-Engineering :

Decreasing vehicle weight and size by use of more advanced technology is EXPENSIVE !!

Performance optimization is out !!

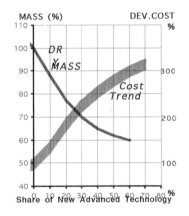

### Conclusion:

Design vehicle with existing technology unless practical reasons (too large dimensions, special technology not available) prevent this

**Figure 4.** Reduction of RLV dry mass using advanced technologies demands high development cost

**References:**
1.   Koelle, D.E.: TRANSCOST 6.2 - Statistical-Analytical Model for Cost Estimation and Economical Optimization of Space Transportation Systems, Edition 6.2, Oct.1998
2.   Bekey, I.: Why SSTO Rocket Launch Vehicles are now Feasible and Practical- A White Paper, NASA Headquarters., Jan.1994
3.   Koelle, D.E.: Economics of Fully Reusable Launch Systems (SSTO vs. TSTO Vehicles), Paper IAA-96-IAA.1.1.03, 47th International Astronautical Congress, Beijing/China, Oct.1996

# Innovative Breakthroughs to a Reusable STS

**T. Yamanaka,** National Aerospace Laboratory, 44-1 Jindaiji-Higashi-Cho, Chofu, Tokyo 182-8522, Japan

e-mail: tatamont@propel.ne.jp

**Abstract**

Markets for fully-expendable/partially-expendable rockets including Two-Stage-To-Orbit (TSTO) and for a fully reusable human-rated Space Transportation System (STS) are discussed. Internet-related space business and in-orbit maintenance, the re-supply of consumables, and the replacement of sub-systems are considered as the most important future markets for launchers. The corresponding state-of-the-art technologies of STS are reviewed, pointing out the problems. A new propulsion system, which could lead to a breakthrough for a reusable STS, is proposed.

## 1. Launch Market

### 1.1 Market for Expendable Rockets

Since the early of 1970's, commercial space payloads have increased in number. Most of them were communications satellites in Geosynchronous Earth Orbit (GEO). During the last decade, many ambitious commercial satellite programs have been proposed and some of them are now in their mission orbits. Most of them are communications satellites; however, the constellation concept to cover a global user area by the use of Low Earth Orbit (LEO) satellites greatly increases their number. Satellites for GEO commercial communications, commercial Earth observation, science and government-related space business are also increasing. In this new decade, about a thousand commercial and governmental satellites are forecast to be launched, even if the "so-called Iridium flu" is embarrassing the investment market.

Many rockets are being developed to launch these space payloads, and the lack of sufficient business to sustain so many launch vehicle programs is of concern. All of the currently available launch vehicles are rocket powered multi-stage expendables.

A new driver for the market will be data communications systems that use satellites to transmit signals; these have many advantages over ground-based systems. Norcross [Reference 1] states that the majority of broadband communications will be provided by satellite systems. He supposes more than 400 satellites providing Internet users with low-cost, direct-to-the-home connections that are hundreds of times faster than today's modems. His forecast is based on the comparison between ground-based and space-based systems evaluating ubiquity, economics, performance and competitive diversity. He also

*M. Rycroft (ed.), The Space Transportation Market: Evolution or Revolution?*, 167–175.

notes that, with market analysts expecting satellites to serve 15 to 20 percent of broadband subscribers, there should be room in the market for multiple competing systems to offer customers Internet connections in orbit.

The author forecasts that the world commercial, governmental, and scientific satellites market for 2005-2015, including the above mentioned new space-based digital data communications systems, will number 300 units and generate over US $ 16 billion in revenue, extending the forecast of Frost and Sullivan 1995-2005 [Reference 2]. The Internet business, however, has a 10-month instead of a 10-year outlook. Space business should shorten the period of time required for developments toward such a new market.

### 1.2   Market for a Reusable STS

The US Space Shuttle has developed various potential space payloads since its first flight in 1981. Large space structures stimulated the imagination of scientists and led to various concepts such as space power satellite systems, hotels in space for tourism, space colonies, etc., which were far beyond the capabilities of the Space Shuttle. Large antennas for advanced communications as well as for advanced Earth observations will be promising markets for commercial and governmental purposes. Special space payloads requiring astronaut-tended missions for regular maintenance (e.g. the Hubble Space Telescope), micro-gravity sciences and commercialization have been other aspects of new space markets. The author believes that in-orbit maintenance, re-supply, and replacement of sub-systems will be the most important issues for the future fully reusable human-rated space transportation system (STS). Activities at the International Space Station (ISS) may require the current US Space Shuttle fleet to be expanded.

Because of the very high cost and of the too long turn-around-time of the US Space Shuttle, the newly obtained knowledge and technologies have been restricted to developing a new market. Almost more than three decades have passed since the Space Shuttle technologies were developed. However, state-of-the-art technologies do not show a viable human-rated fully reusable STS which improves on the Space Shuttle for new space markets; innovative breakthroughs to a reusable STS are required.

## 2.   State-of-the-art Technologies of STS

### 2.1   Expendable Rockets and TSTO

Commercial, governmental and scientific satellites forecast for 2000-2005 will be the markets of all-rocket launchers. All-rocket launchers include various configurations such as multi-stage expendables, partially expendables, turbine-based Two-Stage-To-Orbit (TSTO) and fully reusable Single-Stage-To-Orbit (SSTO) (e.g. VentureStar). The most important factors in terms of assessing the launch vehicle for commercial satellite customers are the reliability of a launch vehicle and the cost. Launch success is the most critical. Consistently on-schedule launches of satellites into the correct orbits are a basic requirement because insurance does not cover lost revenues due to launch delays.

The reliability of launchers depends deeply on in-orbit weight fraction to the gross lift-off weight (GLOW) from the technological point of view for system design, as well as on manufacturing and operational maturity. The maturity of the vehicle is revealed in its flight record. Fully expendable rocket vehicles have a propellant weight fraction to the GLOW of about 80 %. The residual weight is composed of structures, power plants, avionics, payloads, and ballast. Reliability is based on safety design, which consists of the strength margin of structures, multiple warning systems, and system redundancy. The increase of reliability, consequently, increases the dry weight fraction of the vehicle, which, simultaneously, reduces the payloads mass, with penalty of high prices for the payloads.

An all-rocket SSTO carries a more than 90 % propellant weight fraction to the GLOW, which requires advanced composite cryogenic tank structures instead of metals such as aluminum alloys. If 10-20 tonnes is assumed for LEO payloads, the vehicle size of an all-rocket SSTO becomes huge, over 2,000 tonnes. TSTO is a technologically acceptable concept because the required propellant weight fraction is much lower than for the all-rocket SSTO. The propellant weight fraction of the all-rocket TSTO (fly back booster + expendable launch vehicle) is essentially of the same order as for fully expendable multi-stage rocket vehicles. However, the launch cost will be reduced to about one third of that for fully expendable rocket vehicles [Reference 3]. If air-breathing power plants such as turbo, Rocket Based Combined Cycle (RBCC), ram and scramjet are integrated to the fly back booster, the propellant weight fractions are about 10 % lower than for an all-rocket TSTO. These values, however, do not support heavy payloads because of the limitation of gross take-off weight (GTOW) [Reference 4] and of a large enough reliability for human-rated design. Therefore, the TSTO vehicles do not improve the launch cost so much.

If the commercial, governmental and scientific satellites market size increases much more than the previously stated forecast in future, expendable rockets will have the costs of a partially reusable TSTO, by the logarithmic diminution law of mass production and by mature manufacturing and operations.

## 2.2    Airbreathing Engine Powered SSTO

The largest market for a reusable STS is the current payload of the US Space Shuttle and will naturally be those of the ISS. Therefore, the new reusable vehicle should be a human-rated SSTO. Hypersonic technologies have been studied, specifically concerning high-speed propulsion, propulsion-airframe integration, structures, materials and thermal managing concepts, toward a viable SSTO, since the US National Aerospace Plane Program (NASP) was initiated in the middle of 1980's. Viable and practical air-breathing, hypersonic propulsion systems for a fully reusable human-rated SSTO have been a primary focus of those efforts for fifteen years. Two typical concepts are currently studied for that purpose, i.e., turbine-based low-speed propulsion and dual mode scramjet high-speed propulsion [Reference 5] and RBCC powered VTHL (Vertical Take-off Horizontal Landing) vehicles [Reference 6].

The combination of turbine-based low-speed propulsion and dual mode scramjet high-speed propulsion is called the CTSC engine. A human-rated fully reusable SSTO concept powered by the CTSC engine, serving the ISS operation, has to carry about 67-70 % propellant weight fractions to the GTOW of 450 tons, while slush hydrogen is used for fuel [Reference 5]. If we hope for a much greater reliability of the vehicles, two orders more than that of the current US Space Shuttle, we need much lower propellant weight fractions.

The CTSC engine system has very high specific impulses (Isp) in each region of flight velocity from the takeoff to the very high flight Mach number, due to the use of a turbine engine, a pure dual mode scramjet and a liquid-oxygen (LOX) augmented scramjet, respectively. All modes of the CTSC engine are based on the pure Brayton cycle for the working fluid of air, which requires a very high dynamic pressure (96 kPa) flight path. The CTSC has many technological problems to develop. First, the high dynamic pressure flight path produces higher drag on the vehicle, which consequently induces a lower effective Isp. Secondly, the CTSC powered vehicle requires multiple air flow paths, and very sophisticated propulsion-airframe integration technology. Further, the pure Brayton cycle requires variable geometry for the supersonic diffusers. Either of multiple airflow paths and variable geometry of engine have severe thermal management problems in the hypersonic flight region. Thirdly, there is a limit to the maximum flight velocity for a pure Brayton cycle engine; the limit under flight conditions is Mach number 18.

A rocket-based combined cycle (RBCC) powered SSTO is currently believed to be promising, because of its simpler propulsion-airframe integration due to one air flow-path. The Strutjet powered VLHL SSTO concept shows 80%

of total propellant fraction to the GLOW of 747 tons [Reference 6]. Considering the thermal management problem, the transition flight Mach number from scramjet to rocket mode was conservatively designed between 9 and 12. This vehicle concept is believed to be most promising for next generation STS to the ISS; however, a large enough reliability requires a much lower total propellant fraction to the GLOW. The author supposes that the reason is not only due to its concept of sequential operation of the Strutjet and ram/scramjet but to the limit of maximum flight velocity because of the pure Brayton cycle, as previously stated.

## 3.    ARCC Engine Powered SSTO

The most important technological issue is the propulsion system required for a viable and practical fully reusable human-rated SSTO. The author has proposed a new propulsion engine concept [Reference 7]. The concept is basically a kind of RBCC engines; however, its geometrical configuration and operation are different from that. The propulsion system is, therefore, called an airbreather and rocket combined cycle (ARCC) engine.

### 3.1   ARCC Engine

An airbreather engine for SSTO must be evaluated by its performance along the flight corridors together with the vehicle configuration, because the engine characteristics are closely related to the vehicle configuration. Fig. 1 shows a schematic of the reference vehicle.

**Figure 1.** Schematic diagram of the reference configuration

The author believes that the propulsion system concept should be simple. Therefore, variable geometry should be used at least for the air intake, engine internal flows such as the diffuser and air combustor, and nozzle. He believes also that airbreathing engines have to work at higher flight velocities with a higher effective Isp, superior to those of pure rocket engines.

Fig. 2 shows two sided schematics of the propulsion concept. A rocket engine is contained in each strut as the Strutjet [Reference 8], and plays various roles in the ARCC engine. The first is as rocket engines themselves. The second is as ejectors to draw air in the forward part of mixing between the rocket exhaust gases and the incoming air flow, from low subsonic flight velocities to hypersonic flight velocities exceeding the rocket exhaust velocity, i.e., to about $V_\infty \sim 4$ km/s, which produces thrust augmentation. The third is as flame holders for air/fuel combustion in the air combustor. The fourth is as hydrogen fuel and oxygen injectors to the air combustor by using lower and higher o/f (mass flow rate of oxygen to mass flow rate of fuel of the rocket) compared with those of conventional LOX/LH2 rocket engines. The fifth is as pressure controller for the air combustor, such as the role that the rocket exhaust gas plays of adjustable supersonic diffuser for supersonic combustion to keep the combustion pressure within reasonable values without variable geometry. The fifth function is a unique feature of the ARCC engine.

If one wants to maximize the performance of an airbreather for very high flight Mach numbers, area contraction ratios of the forward air flow passage in front of the combustor must be small; otherwise various internal flow choking phenomena will be induced. The smaller contraction ratios need a lower pressure combustor, which requires a longer combustor to enable both mixing and combustion. If higher contraction ratios (supersonic diffusers) are designed for those parts, the airbreathing engine does not work at very high flight velocities. The exhaust gases of the rockets can increase the pressure of the combustor via mixing processes, which simultaneously extends the maximum flight Mach number of the ARCC scramjet mode whilst avoiding hypersonic air dissociation effects. This is the fifth function of the rockets in the ARCC engine as mentioned above.

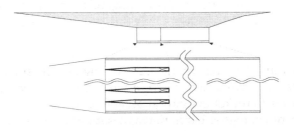

**Figure 2.** Schematic of the ARCC propulsion concept

Thermal management problems are very severe for the Strutjet type RBCC propulsion system, as Siebenhaar has stated [Reference 8]. In order to avoid such problems, the airbreather cycle of their Strutjet is limited to a low flight Mach number of about 8. Even if the airbreather cycle of the Strutjet is limited to lower flight Mach numbers, the strut and airbreathing combustion chamber wall structures need to be cooled by hydrogen fuel through wall jackets like those of conventional LOX/LH2 rocket engines, because of high recovery temperatures on the walls. Supposing a material with high thermal conductivity and high stiffness, such as copper-based composites, for the wall structures, the author considers the value of transpiration cooling by means of hydrogen fuel to the walls of the ARCC engine.

## 3.2  *Performance of the ARCC Engine*

Two kinds of computer programs have been developed to estimate an ARCC engine performance. One is for lower flight velocities from takeoff/liftoff to about $M_\infty \sim 3$, in which region, the engine structures (except air/fuel combustion) are at the limits of high temperature resistance without active cooling. Transpiration cooling is applied only to air and fuel combustion parts in the lower flight velocity region and to all parts of the engine internal surfaces (except ramp and nozzle) in the higher velocity region. Preliminary flight estimates of the reference vehicle powered by an AECC engine showed a better vehicle mass fraction performance through the lower flight-dynamic-pressure paths such as 1/5 atmospheric pressure. The minimum dynamic pressure of the higher flight velocity region was limited to 1/3 atmospheric pressure.

Numerical results obtained by the computer programs are shown in Figs. 3-a and 3-b, which present specific impulses versus flight velocity of the lower and higher velocity regions, respectively. Here, pc is the pressure of the rocket engine combustion chamber, $b_r$ is the width of rocket engine, b is the pitch of struts and $b_r/b$ is the blockage ratio. It is to be noted here that a higher blockage ratio deteriorates Isp. The figures show excellent Isp values from takeoff/liftoff velocities to very high velocities (more than 6 km/s). Thrust/(frontal air intake area) is an important parameter for the vehicle designer; this is also shown to be very good.

The most important features of the ARCC engine are summarized as follows. First, a lower pc gives a higher Isp in the lower flight velocity region, which is closely related to rocket engine ejector effects by means of pc control. Second, a higher pc gives a higher Isp in the higher flight velocity region, which is related to the supersonic diffuser mechanism without variable geometry. Third, a lower o/f means that real combined cycles (parallel operation) of rocket propulsion and the Brayton cycle are possible because of smaller transpiration cooling fuels in the lower flight velocity region. The higher o/f in the higher flight velocity region means that increased transpiration cooling fuels are burnt with air in the combustor. These features are different from those of a RBCC engine such as the Strutjet. The author supposes the propellant weight fraction of the ARCC powered SSTO will be below the minimum of the CTSC. A computer program estimating an ARCC powered SSTO reference vehicle is now being prepared by the NAL ARCC research group. The facilities to verify performance and the characteristics of the ARCC by ground-based experiments are limited. Flight experimental vehicles are consequently necessary.

### References

1.  Norcross, R. P.: Satellites: The Strategic High Ground, *Scientific American*, *Vol. 281*, pp. 86-87, October 1999
2.  Smith, B. A.: New Launchers Seek Commercial Market Share, *Aviation Week*, *Vol. 151*, pp. 50-52, December 13, 1999
3.  Kostromin, S. F.: *Cost Effectiveness Estimates of the Partially Reusable Launchers Family with Uniform Components*, AIAA 99-4887, presented at the 9th International Space Planes and Hypersonic Systems and Technologies Conference, Norfolk, VA, USA, November 1-5, 1990
4.  Olds, J. R., et al.: Stargazer: *A TSTO Bantax-X Vehicle Concept Utilizing Rocket-Based Combined-Cycle Propulsion*, AIAA 99-4888, ibid.
5.  Moses, P. L., et al.: *An Airbreathing Launch Vehicle Design with Turbine-Based Low-Speed Propulsion and Dual Mode Scramjet High-Speed Propulsion*, AIAA 99-4948, ibid.
6.  Stemler, J. N., et al.: *Assessment of RBCC-Powered VTHL SSTO Vehicles*, AIAA 99-4947, ibid.
7.  Yamanaka, T., et al.: *Airbreather/Rocket Combined Cycle (ARCC) Engine for Spaceplanes*, AIAA 99-4812, ibid.
8.  Siebenhaar, A., et al. :*The Strutjet Engine: The Overlook Option for the Space Launch*, AIAA 95-3124, presented at the 31st AIAA/ASME/SAE/ASEE Joint propulsion Conference and Exhibit, San Diego, CA, USA, July 10-12, 1995

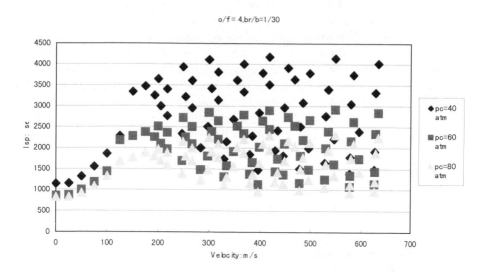

a) Lower flight velocity region

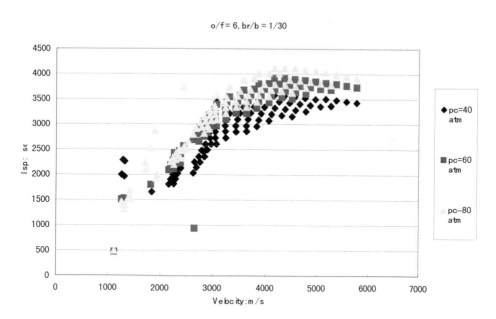

b) Higher flight velocity region

**Figure 3.** Specific impulse (Isp) of ARCC engine (br/b=1/30)

# A Japanese Road Map for
# Future Space Transportation System

S. Nomura, National Space Development Agency of Japan (NASDA), 2-4-1 Hamamatsu
-cho, Minato-ku, Tokyo 105-8060, Japan

e-mail: Nomura.Shigeaki@nasda.go.jp

### Abstract

It has long been proposed that a reliable space transportation system (STS) should
be developed in Japan as in many other countries. Recently this becomes inevitable, due
to successive launch failures of rockets which occurred in the last three years in Japan.
Consequently an ad hoc Committee on Future Space Transportation System has been
established by the Science and Technology Agency of Japan. Extensive investigations
were conducted about the essential features of a future reusable launch vehicle (RLV)
and the technological maturity for vehicle core systems. In conclusion, a fully reusable
two stage to orbit (TSTO) spaceplane system equipped with air-breathing engines for its
first stage was selected as a target RLV system to be developed in Japan around the mid
2010's. A road map for research and development towards a future RLV was shown
where a well-mixed space infrastructure of RLV and expendable launch vehicle (ELV)
was proposed. This paper overviews the technological background to develop a future
RLV in Japan and outlines the Committee's final report issued in June 2000.

## 1.    Introduction

In many countries great efforts have been made to develop a RLV which
will provide a much better STS than the present one in terms of reliability,
safety, operational capability and costs for development, manufacture, and
operation. Though many kinds of RLV vehicles of the single stage to orbit
(SSTO) type and TSTO type were proposed, none of them has yet succeeded.
Now several RLV projects are on going and many new concepts have been
proposed by the private sector; all present projects have been confronted with
difficulties to a certain extent, whether of an economic, political or technological
nature. Development costs of future RLV systems are so high as sometimes to
cause a political problem for national development organizations. From the
economic standpoint the high development costs invested should be recovered
and an assurance of drastic operational cost reduction should be, therefore, an
indispensable issue, especially for the private sector. On the other hand,
technological maturity is the most important aspect, though it was problematic
in some previous projects; a technological challenge was sometimes confused
with technological immaturity.

For a long time it has been proposed to develop a reliable STS in Japan
[References. 1, 2] as in many other countries, but it becomes inevitable due to
recent successive failures of rocket launchings in last three years. Under such a
situation an ad hoc Committee on Future Space Transportation System was

M. Rycroft (ed.), The Space Transportation Market: Evolution or Revolution?, 177–185.

established in the Science and Technology Agency of Japan (STA) in April 1999. In this Committee a future target of a Japanese RLV and a road map for research and development toward a future RLV were investigated extensively. A TSTO spaceplane equipped with air-turbo ramjet engines for its first stage and reusable rocket engines for the upper stage was proposed as a target RLV system to be developed around the mid 2010's in Japan. A road map towards this future system was also shown where a well-mixed system of ELV and RLV was proposed as a future space infrastructure. The report of the Committee was issued temporally in October 1999 and finally in May 2000 [Reference 3]; international cooperation was strongly emphasized. The proposed TSTO system is not new [Reference 4]. Systems named Saenger had been studied in Germany, and the most recent one conducted by Deutsche Aerospace [Reference 5] was interrupted mainly due to financial difficulties. Concerning the Japanese technological base necessary for a TSTO system, the most important studies for a TSTO system have proceeded in the area of engine systems and vehicle systems. Extensive investigations have been made on an expander cycle air-turbo ramjet engine (called ATREX, [Reference 6]) and the vehicle system development program of the H-II Orbiting Plane (named HOPE, [Reference 7]).

This paper overviews the Japanese technological base for RLV development focused on engine systems and vehicle systems, and explains the road map of STS development towards a future RLV as described in the final report of the Committee.

## 2.    Technology Development Programs Towards a RLV

Technological developments necessary to realize a RLV could be classified into two categories: (1) engine system technology developments, (2) vehicle system technology developments, including flight verification technologies. The activities in these key technology development areas in Japan are outlined here.

### 2.1    Air-breathing Engine System Technology

Two typical development activities are briefly overviewed here. Besides these, there are other important activities concerning the HYPR Project [Reference  8] for a high-speed turbo-ramjet engine by National Aerospace Laboratory (NAL), and long life rocket engine studies by NASDA [Reference 9] and NAL [Reference. 10].

An expander cycle air-turbo ramjet engine called ATREX which is designed to work up to Mach 6 has been developed for ten years by ISAS as a candidate propulsion system for a fly back booster for a future TSTO spaceplane. An ATREX engine system (Fig. 1) consists of an automatically controlled air intake, a pre-cooler system with defrost devices, three-staged fans with tip turbines on fan peripheral tips, a regeneratively cooled combustion chamber, and a variable configuration nozzle. ATREX engine combustion tests under sea level static conditions have been conducted successfully. The test engine of total length of 5 m and fan inlet diameter of 0.3 m has been equipped with metallic tip turbines, a bell mouth intake and an exhaust convergent nozzle. The flight engine will be, however, equipped with a variable configuration air intake, a plug nozzle, and tip turbines of advanced carbon/carbon composite materials which are under development. Wind tunnel tests of an air intake equipped with automatic control systems have been conducted extensively by using several tunnels of ISAS, NASA Glenn Research Center and ONERA. Those test results showed that the design performance was almost reached and that the automatic control system worked well to recover from unstarted conditions.

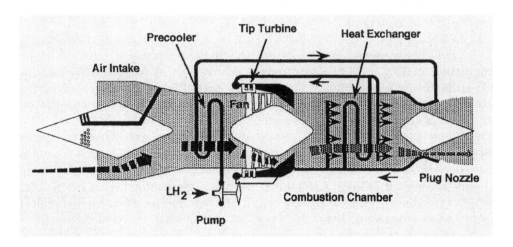

**Figure 1.** ATREX engine system

A proposed system concept of an ATREX engine flight test [Reference 11] is as follows; (1) A flying test bed will be self-powered by an ATREX engine with 30 cm fan diameter under an air frame, (2) Taking-off horizontally with a take-off assist booster, flying forward, and being recovered from the sea by a parachute, (3) Vehicle characteristics; total length of 12.5 m, delta wing span of 2.7 m, gross take-off weight of 1.6 tonne with fuel of 0.2 tonne, and (4) The maximum speed of Mach number 6 will be reached at about 30 km altitude.

At NAL a supersonic combustion ramjet (Scramjet) engine has been developed. Combustion tests of sub-scale scramjet engines [Reference 12] of 2.1 m long with a front area of 20 by 25 cm have been conducted under flight conditions for Mach numbers of 4, 6 and 8. Combustion tests conducted in the Ram Jet Engine Test Facility at NAL more than 130 times showed a net positive thrust for these Mach numbers. Tests at higher velocities around 4 - 7 km/s under actual flight conditions of a SSTO are being conducted in the High Enthalpy Shock Tunnel constructed recently at NAL.

## 2.2 Vehicle System Technology: Flight Experiment Series of the HOPE Program

The HOPE program consists of many precursor experimental flight projects of small vehicles and flight experiment projects of relatively large scale vehicles flying at supersonic and orbital reentry speeds. The already completed flight experiments in the precursor series are: (1) Orbital Reentry Experiment (OREX, [Reference 13]) conducted successfully in 1994, (2) Hypersonic Flight Experiment (HYFLEX, [Reference 14]) conducted successfully in 1996, and (3) Automatic Landing Flight Experiment (ALFLEX, [Reference 15]) conducted successfully at Woomera, Australia in 1996.

The final program in the precursor flight experiments series will be conducted in 2001 - 2002. This program [Reference 16] consists of two flight experiment phases as follows: (1) Phase 1: A HOPE configuration vehicle of about quarter scale with a small jet engine will take off, fly at subsonic speed, and land horizontally at a runway on Christmas Island, and  (2) Phase 2: The same scale vehicle will be dropped from a balloon to make transonic speed flight experiments in Sweden in cooperation with CNES. The main HOPE program consists of two successive experimental projects of several supersonic flights and an orbital flight as shown in Fig. 2. Its configuration has a total length of 15.2 m, span of 9.7 m, and height of 4.8 m. A supersonic HOPE vehicle of full scale, is equipped with rocket engines to take-off vertically from a launch pad which will be constructed on Christmas Island. After supersonic flight tests the vehicle will fly back to the same location to land horizontally on a runway.

The HOPE orbital reentry vehicle is a reinforced version of a supersonic vehicle using suitable thermal protection systems and new materials. It will be launched by the first stage of a standard type H-IIA rocket of NASDA: under development to be launched at the beginning of 2001 [Reference 17]. Rocket engines within the vehicle will be ignited after separation from the H-IIA to insert it into low Earth orbit (LEO). An orbital reentry flight experiment will take place after one revolution and the vehicle will land horizontally on a runway on Christmas Island.  The HOPE program is planned to reach the final stage of supersonic flight experiments in 2003FY and an orbital reentry flight experiment in 2006FY.

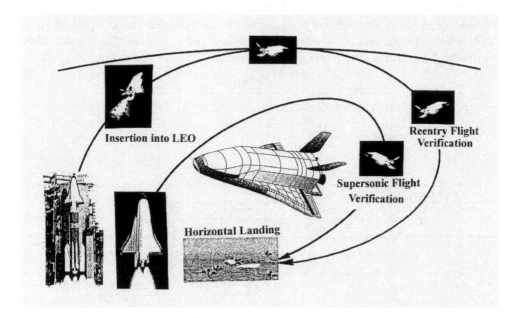

**Figure 2.** HOPE flight profiles of two experimental vehicles

## 3.    Development Scenario of RLV in Japan

The ad hoc Committee on Future Space Transportation System was established in STA in April 1999 as described in Section 1. Extensive investigation in the Committee was especially focused on a target year of the operational RLV development, essential features of the RLV at that time, and expected levels of technological maturity for the vehicle's core systems. Based on these studies, it was concluded that an operational RLV should be developed around the mid 2010's and that a target RLV system in Japan should be a TSTO spaceplane equipped with air-turbo ramjet engines for its first stage and reusable rocket engines for the upper stage. A road map (see Fig. 3) towards this future system was also shown where a well-mixed system of ELV and RLV was proposed as a future space infrastructure. The final report of the Committee was issued in June 2000; in this an international cooperation was strongly emphasized.

The development scenario of a future RLV in Japan consists of three research and development (R-D) phases. In phase 1 (see Fig. 4) R-D will be focused on engine system developments of an air-breather and a reusable rocket and on vehicle system developments. As for an air-breather, it will take about 5 years to develop an ATREX engine with a fan diameter of 40 cm. It will be supplied to a flying test vehicle to evaluate engine performance at up to Mach 6. A reusable rocket engine of 10 tonnes thrust revel will be developed after technology investigations. Vehicle system technologies in supersonic and orbital reentry flight regions will be evaluated by successive flight experiments in the HOPE program. The experiences and technologies accumulated in the HOPE program will be reflected mainly in the second stage developments of a TSTO vehicle.

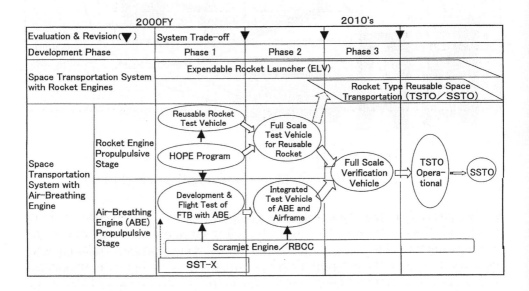

**Figure 3.** Road map towards future RLV development in Japan

In phase 2 flight test vehicles of large scale will be developed for both stages, and these will be tested separately. Two sets of ATREX engines with 60 – 80 cm fan diameter will be installed in the first stage vehicle of lift-off weight 20 tonnes which will fly at up to Mach 6 at 30 km altitude and will fly back to the take-off runway. The second stage vehicle of 120 tonnes equipped with the developed reusable engine of 100 tonnes thrust level will be launched and inserted into LEO by an expendable rocket. It will fly back after an orbital reentry test to land horizontally on a runway.

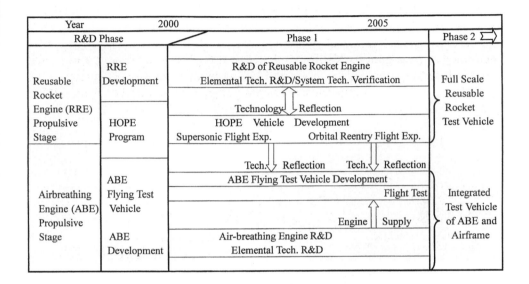

**Figure 4.** Technology development in phase 1

In phase 3 a full scale verification vehicle of a TSTO system will be developed. It will have payload capability of 8 tonnes into LEO. The first stage will be equipped with 6 sets of ATREX engines with fan diameter of 120 cm and the second stage is the same as developed in phase 2. Weight characteristics, figures obtained from feasibility studies, are as follows: dry weight of 89 tonnes for the first stage, 21 tonnes for the second stage, and lift-off weight of 270 tonnes for total vehicle (140 tonnes for the first stage and 130 tonnes for the second stage). After finishing three R-D phases an operational TSTO vehicle (see Fig. 5) could be constructed and would appear as a Japanese RLV after the mid 2010's. It was pointed out in the report that the successful development of a TSTO spaceplane will give a chance to start manned space transportation in Japan.

In parallel with these developments and operations of a RLV system, ELV vehicles should be developed and operated continuously to supply better STS for wide demand, such as transportation to GEO, to deep space, to another planet and so forth. It was also implied by the report that, including a reusable orbit transfer vehicle, a well-mixed system of ELV and RLV should be constructed as the basis for a future space infrastructure.

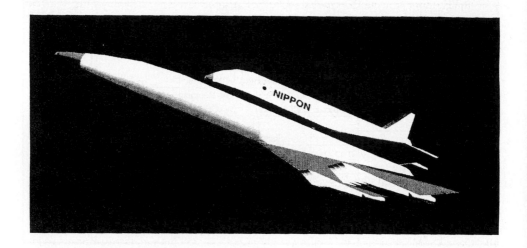

**Figure 5.** Image of TSTO as a target RLV system in Japan

## 4.     Concluding Remarks

A Japanese road map for a future space transportation system and a target RLV system have been presented, where a well-mixed space infrastructure of ELV and RLV was proposed. The development scenario of a future space transportation system consisted of three R-D phases was explained in some detail. These were decided by the ad hoc Committee established in STA. It proposed a fully reusable two stage to orbit spaceplane system equipped with air-breathing engines for its first stage  as a target RLV  to be developed around the mid 2020's in Japan. The final report of the Committee was issued June 2000: an international cooperation was strongly emphasized there.

**References**
1.  Nomura, S.: *Japanese Activities for Future Space Transportation System*, IAF-98-V.3.01, 1998
2.  Nomura, S.: *Japanese Activities for Space Transportation System Beyond 2000*, Proceedings of the AAAS/AAS Symposium "Space Access & Utilization Beyond 2000", April, 2000
3.  Final Report on the Committee of Fully Reusable Transportation System, Science and Technology Agency, June, 2000
4.  Tanatsugu, N. et al.: A Study on Two-Stage Launcher with Air-Breathing Propulsion, Space Exploitation and Utilization, Vol. 60, *Advances in the Astronautical Science*, pp. 365-381, 1986
5.  Koelle, D. E. et al.: *SAENGER Space Transportation System - Progress Report 1990*, IAF-90-175, 1990
6.  Sato, T. et al.: *Development Study on ATREX Engine System*, IAF Paper 99-S.5.04, 1999
7.  Fukui, T. et al.: *Present Status of H-II Orbiting Plane-Experimental (HOPE-X) Development*, Proc. Space Technology and Applications International Forum, pp. 937-942, January 1998
8.  Nimura, N. et al.: *Experimental Approach to the HYPR Mach 5 Propulsion System*, AIAA Paper 98-3277, 1998
9.  Taniguchi, H. et al.: *Rocket SSTO Concept Study*, International Symposium on Space Technology and Sci. (ISTS) Paper 98-o-1-04V, also IAF Paper 97-V.3.05, 1997
10. Nosaka, M. et al.: *Tribo-Characteristics of Cryogenic Hybrid Ceramic Ball Bearings for Rocket Turbopumps: Self-lubricating Performance*, Trib. Trans., Vol. 40, No. 1, pp. 21-30, 1997
11. Sato, T. et al.: *Development Study for ATREX Engine Flight Test*, ISTS Paper 98-a-1-29, 1998
12. Chinzei, N. et al.: *Sub-Scale Scramjet Engine Tests at NAL-KRC*, ISTS Paper 98-a-1-26V, 1998
13. Inouye, Y.: *OREX Flight - Quick Report and Lesson Learned*, Proc. of 2nd European Symposium on Aerothermodynamics for Space Vehicles, pp. 271-278, 1994
14. Shirouzu, M.: *On the Hypersonic Flight Experiment (HYFLEX) for the Development of HOPE*, AIAA Paper 93-5080, 1993
15. Miyazawa, Y. et al.: *Flight Testing of ALFLEX Guidance, Navigation and Control System*, ICAS Paper 98-1.1.3, 1998
16. Yanagihara. M. et al.: *Simulation Analysis of the HOPE Demonstrator*, AIAA Paper 99-4875, 1999
17. Hirata, K. et al.: *H-IIA System Design*, 21st ISTS Paper 98-f-07, 1998, also Watanabe, A. et al., IAF-97-V.1.04, 1997

# Report on Panel Discussion 3:

# Reusables and Expendables in the Future Launch Market: What Will be the (Right) Mix? And Where is Technology Heading?

**C. Lo, M. Munoz Fernandez, W. Soh,** International Space University, Strasbourg Central Campus, Parc d'Innovation, Boulevard Gonthier d'Andernach, 67400 Illkirch-Graffenstaden, France

e-mail: lo@mss.isunet.edu, munoz-fernandez@mss.isunet.edu, soh@mss.isunet.edu

**Panel Chair : S. Nomura,** NASDA, Japan

**Panel Members:**

**Part I:**

**W. Fletcher,** NASA Kennedy Space Center, USA
**T. Ito,** NASDA, Japan
**T. Lacefield,** Lockheed Martin Aero-Palmdale, USA
**R. Kohrs,** Kistler Corporation, USA
**T. Morrison,** Boeing Delta Launch Services, USA

**Part II:**

**R. Lindberg,** Orbital Sciences Corporation, USA
**M. Eymard,** CNES, France
**M. Caporicci,** ESA, France
**D. Koelle,** TransCostSystems, Germany
**T. Yamanaka,** National Aerospace Laboratory, Japan

The panel discussion involved participants from government agencies and private companies. Part I of the session was more technically oriented, while Part II focused more on futuristic topics concerning both the market environment and technology issues for the launcher business.

Part I of the panel discussion focused mostly on the implementation of the reusable launch technologies. The large number of questions posed by the audience reflected the high level of interest in the subject. In answering one of the selected questions from the audience, **T. Lacefield** explained that the

187

*M. Rycroft (ed.), The Space Transportation Market: Evolution or Revolution?, 187–189.*
© 2000 *Kluwer Academic Publishers. Printed in the Netherlands.*

possibility of using the Venture Star's external payload cabin for carrying a human crew was being explored.

Several questions were asked of **R. Kohrs** of Kistler Corporation. In reply, he clarified that the legal and political issues of the K-1 operations were under the regulation of the Australian Government. On a more technical note, he detailed that the parachutes and air bags used for the K-1 landings have a design lifetime of 6 flights. From the start of operations, hopefully next year, Kistler plans to schedule 4, 7 and 13 launches in the first, second and third years, respectively. The K-1 could carry 1 tonne of payload to the International Space Station.

When asked of the future of Delta II and Sea Launch when Delta IV comes into operation, **T. Morrison** replied that, to his knowledge, both Delta II and Sea Launch will be kept in operation independently in order to maintain a range of launch capabilities within Boeing.

Part II of the panel discussion was held after the presentation of all the papers in the session. Addressing the opening question on cost engineering, **D. Koelle** confirmed that the TransCostSystems model does take into account weight and complexity when performing a cost sensitivity analysis for a launch vehicle.

In answer to a question regarding the comparison between the X-34 and the Pegasus vehicles, **R. Lindberg** commented that the two launchers use the same air launch aircraft, and have the same avionics and guidance systems. However, the hardware on the X-34 is different from that of the Pegasus, considering that the Pegasus is expendable while the X-34 is reusable. In case of the need to abort a mission, the X-34 flight termination system performs a pneumatic override on the hydraulics, causing the wing flaps to move so that the vehicle becomes non-lifting and rolls, so terminating its ballistic trajectory.

There was a consensus among the panel members that reusable launch vehicle technologies will eventually replace expendable ones. **M. Eymard** commented that the expendable launch providers must prepare themselves for the advent of the commercial reusable launcher market. Nevertheless, the initial development on reusable vehicles covers only low Earth orbit (LEO), while the commercial market demand is still very much concentrated on launches to the geostationary transfer orbit (GTO). In the short term, Arianespace will not be developing reusable launch vehicles (RLV). But, for the future, the company must take the up-and-coming reusable LEO launch market into consideration, in order to keep up with the market trends. But, as **M. Caporicci** pointed out, this will depend very much on the progress of the market for launching LEO satellite constellations. The complexity of missions to LEO must also be compared with that of missions to GTO.

In response to a question from the audience, **M. Caporicci** explained that the European transfer vehicle to the International Space Station will only carry fluids and propellant, and will not cater for crewed operations. **M. Eymard** asserted that the European participation in the crew return vehicle (CRV) as a complement to the Space Shuttle will hopefully reduce costs.

On the topic of the Siamese two stage to orbit configuration, with two identical vehicles forming the two stages, **M. Caporicci** commented that, while it may be cheaper to develop, it may not be technically optimised. Problems may also be encountered in the separation of the two stages.

A member of the audience questioned how the lack of a production schedule of reusable vehicles like the X-34 could lower the launch costs or retain skills among the workforce. **R. Lindberg** stated that the X-34 is not a prototype of a commercial RLV; instead, it is a testbed for specific research purposes. Extending the technology to create a RLV would be a considerable challenge, which may benefit from drawing on much experience in the aviation field.

In considering the potential commercialisation of the RLV in the future, a question was raised in view of the fact that the entire satellite launch market could be satisfied with 5 to 6 RLVs. Competition in the launch market would thus seem to be restricted. A member of the audience commented that the market could expand to cover the carriage of passengers, which would maintain the market volume, while industrial customers could expect more reasonable launch costs. **M. Eymard** added that the transition from expendable launch vehicles (ELVs) to RLVs would be introduced over a long period of time, since RLVs require very different scheduling. Operations and maintenance experience will be an important cost factor for the RLV; an analogy can be drawn here to the Concorde supersonic aircraft. **R. Lindberg** added that the industrial demand for access to space provided by more than one provider would also ensure the continuation of a competitive launcher market.

Finally, a question was raised regarding the noise generated by RLVs while travelling in the atmosphere. As for existing supersonic aircraft, RLVs would fly at subsonic speeds over populated areas, and only accelerate above the high seas.

# Session 4

# How will the Legal Frameworks Need to Evolve?

Session Chair:

**A. Kerrest,** Faculté de Droit de Bretagne Occidentale, France

# How will the Legal Frameworks Need to Evolve?
# Definitions and Legal Issues

**A. Kerrest,** University of Western Brittany, 12 rue de Kergoat, 29200 Brest, France

e-mail: Armel.Kerrest@univ-brest.fr

### Abstract
When the fundamental rules of space law were adopted by UN resolutions and treaties, space transportation as it is currently foreseen was science fiction. Nevertheless the principles governing space activities are in force; they can adapt to suit the new activities. A general consensus exists within the COPUOS to keep the present rules, but in some cases they should be complemented or improved.

This paper considers the legal consequences of the current technical and economic evolutions of space transportation.

## 1. Introduction

Outer space is not and has never been a space without law. Strictly speaking, law does not apply to a space; it applies to human activities taking place there. As far as law is concerned, Outer Space is a framework for human activities. This is the reason why human law in general, either international or national, applies to human activities taking place in Outer Space[1]. This does not mean that the special characteristics of outer space have no consequences to its legal regime. Current space law is very much influenced by three main characteristics of Outer Space and Outer Space activities: the international domain, the considerable role of States, and the danger and hazardous character of the activities conducted in Outer Space [Reference 1].

From the beginning, Outer Space was recognised as an international domain. Given the fact that air space is national, the international nature of Outer Space was a necessity in order to recognise a freedom of use. Law applies in outer space as it does on the high seas: there is no States' territorial jurisdiction but only personal jurisdiction. We are in the same situation as out at sea: international law applies, and so does national law given the possible application of rules of conflict of law. It is not necessary to create a complete

---

[1] To take an example : rules compelling the judge to take a decision as article 4 of the *Code Civil*, which make an obligation for the judge to decide despite possible imprecision or lack of clarity of the law should apply for outer space activity as every where else. Using the usual "terrestrial" legal principles, judges can decide any case involving space activities. Where there is a special "space law" rule, he or she will apply it *as "lex specialis"*; where there is none, the usual law, *"lex generalis"*, applies.

*M. Rycroft (ed.), The Space Transportation Market: Evolution or Revolution?*, 193–200.

new *"corpus juris spatialis"*; the usual international and national laws already apply according to a personal State's jurisdiction [Reference 2]. In some cases it may be useful to make special space law rules to take into consideration the special nature of Outer Space.

At the beginning of the space era, only States (and not private entities) were involved in space activities. The resolutions and the treaties setting the basis of space law were based on a compromise between the two space powers of the time: private entities may act in Outer Space but the States are responsible for any "national activity in outer space"..."whether such activities are carried on by governmental agencies or by non-governmental entities". If we consider the fact that Launching States are liable if any damage occurs, and that the State of registration" shall retain jurisdiction and control over such object, and over any personnel thereof, while in outer space or on a celestial body"[2], we can see how important is the role of States in Outer Space activities.

The third characteristic of space activities as far as space law is concerned is the real or presumed danger. If space activities come close to normal activity, will space law take this evolution into account ?

As for any other international activity, space law may be elaborated on two different levels : the international level and the domestic level. At each there is a different law making process. At the international level are international treaties and customary law accepted by the international community. General fundamental rules have to be set at this level which may be compared to the constitutional rule for national law. We already have some treaties and United Nations General Assembly resolutions. They may be improved by international discussions especially within the United Nations Committee on the Peaceful Uses of Outer Space (COPUOS). It should be very difficult to modify them without endangering the whole system.

At the national level, domestic space law should be adopted. Some States already did so — the United States, Sweden, the United Kingdom, South Africa,

---

[2] *Treaty on Principles Governing the Activities of States in the Exploration and Use of Outer Space, including the Moon and Other Celestial Bodies,* Outer Space Treaty (OST 1967 in United Nations Treaties and principles on Outer Space Commemorative edition United (Nations New York 1999 doc A/AC.105/722 at p. 3); *Convention on International Liability for Damage Caused by Space Objects* (Liability Conv.) 1972 in United Nations Treaties and principles on Outer Space Commemorative edition (United Nations New York 1999 doc A/AC.105/722 at p. 11); *Convention on Registration of Objects Launched into Outer Space* (Registration Conv.) 1975 in United Nations Treaties and principles on Outer Space, Commemorative edition, (United Nations New York 1999 doc A/AC.105/722 at p. 18).

Russia and Australia. Some are currently discussing this issue, France and Germany for instance. This level of law is very important when private activity is involved. As far as only States act in outer space, international law may have been sufficient. It is no more the case if private entities are involved; they are not direct subjects of international law. According to Outer Space treaty article VI, States are responsible for implementation of international law by their nationals (entities conducting "national activities" in Outer Space). If some private persons conduct space activity, a domestic space law is a condition of application of this obligation [Reference 3].

The purpose of these introductory remarks is to consider whether the evolution of space transportation in the foreseeable future is going to change the very basis of space law or if, given some improvements, the current principles may be used as the foundations of a more efficient system.

The exposé is in three parts: the first examines whether technical evolutions in space transportation may jeopardise the basic rules of space law, the second points out the issue of commercialisation of space transportation and its consequences on the current rules, and the third considers the possible change of hazardous activities into more common practices.

## 2.     Space Law and the New Techniques of Space Transportation

### 2.1   *Launching from International Spaces*

When technical advancement made possible the launch of space rockets from platforms located on the high seas, new legal problems were raised. The fact of launching from an international territory modified the whole balance of the system making the territorial criterion void. This criterion is essential; it is specially considered in article V§3 of the liability convention. It is the only one which introduces a purely material element connected with a State's territorial jurisdiction. The other criteria make reference to jurisdical links such as nationality or registration and flag. Those criteria are less reliable, less precise and less unambiguous. The criterion of the "State which launches" is demonstrated by the nationality of a jurisdical person such as the company or the participant to an international consortium. The legal link with facilities is the registration of the platform or the flag of the command ship. As we can see in other fields, at sea for instance, entrepreneurs sometimes choose those links very freely. They may be tempted to make this choice in order to avoid enforcement of regulations and control, which are nevertheless indispensable in such a sensitive field as space activities [Reference 4].

In the case of a launch from a flying aircraft, does the launch begin when the aircraft takes off or when the spacecraft separates from the aircraft carrying it ? In the first analysis the spacecraft should be considered as being formed of two or three stages. The first stage could take off from and land at an airport. The same legal issues would arise as for a spaceplane. If the second hypothesis is accepted, the territory of launch should be the air above the high sea and thus an international space leading to the same problems as highsea launches.

## 2.2  *Reusable Launch Vehicles*

When the main space law treaties were ratified, the only launch technique available was the use of Expendable Launch Vehicles. Therefore, launching, was obviously considered as the central issue for control and liability. Thus, when Reusable Launch Vehicles are used, the main question is not re-entry, which is not specially considered by current space law, but re-launch.

Three issues are at stake here:

• Sovereignty over air space or free use of outer space

• Responsibility and liability for space activity or air rules

• Regulations and control, registration, technical requirements for air and space activities [Reference 5].

The first question is, as often in law matters, a question of definition connected with the essential issue, which is: does space law or air law apply ? Legal issues change according to the technique used. The problems for an Earth to Earth vehicle are not the same than for an Earth to Space one. A single stage to orbit vehicle is not the same as  a multi-stage to orbit vehicle.

For all of them, the question of definition of the liable launching State(s) is to be addressed. According to the Outer Space Treaty (OST) and the liability convention, the launching State(s) change(s) at each launch. It is currently the case with the US Space Shuttle, which for the implementation of international space law is considered as a new vehicle at each new launch. At each new launch, a new set of launching States is involved, and a new registration is done. If re-launches become a day to day practice, the current system will be rather difficult to maintain. The mere fact that such a "spaceplane" can land and take off from an airport would make the State of the territory a liable launching State.

When we consider a plane or "spaceplane" flying an Earth to Earth mission, the problem would be the definition and delimitation of outer space. The question was discussed during the last meeting of the UN COPUOS legal subcommittee. For some States, the issue is not an important topic for the time being. When such a vehicle will be used, it will be necessary to define more precisely what is a space object and thus to delimitate air and outer space. If this "plane" uses outer space, for the purpose of space law, it is a space object with every consequence. If it only uses air space that is not the case. The problem may be avoided if a special status is specified for such vehicles.

The re-entry of a spacecraft poses some legal questions. For the time being space law mainly considers a re-entry for liability purposes[3]. For a reusable vehicle, a re-entry is a normal procedure. First of all the States have to consider this re-entry, as did the USA by amending the Commercial Space Launch Act to introduce this issue and rule on re-entry, and Australia whose Space act deals with re-entry[4]. More difficult is the question of a possible right of "innocent passage" into the air space. For the time being the US Space Shuttle may come back using only international or US air space. When any other State will use a reusable launch vehicle it will have the problem of flying over some other State's territory. International law is not quite clear on this issue. Given the very few precedents, in fact only the single mission of Buran, it would be difficult to argue that an international customary law enables such a passage. It could be seen as a logical consequence of the principle of the free use of outer space for every country, but such a basis for a right of passage is rather weak.

In the case of a multi-stage-to-orbit vehicle the question will arise of the legal status of the different stages. Will the whole vehicle, i.e., the two or three parts, be considered as a spacecraft and thus as a space object to which space law should be applicable? In that case it has to be registered and is subject to space law liability from the beginning of the launch; according to the liability convention the launch begins on Earth. The airport used for take off will be a "facility" making the State of location a launching State. If only the space-going part is considered as a space object, the other parts should be aircraft and the launch will take place in the air. In that case the aircraft should be considered as a launch facility, making the State of registration of the aircraft a liable launching State, i.e., one out of possibly several launching States.

---

[3] Liability Convention and *Agreement on the Rescue of Astronauts, the Return of Astronauts and the Return of Objects Launched into Outer Space,* the "Rescue Agreement" 1968 in United Nations Treaties and principles on Outer Space, Commemorative edition, (United Nations New York 1999 doc A/AC.105/722 at p. 18).

[4] US code title 49 chap 701 n° 70104, Australia Space Activity Act 1998 n°123 1998.

The question of integration of a spacecraft into air navigation is certainly a very important and difficult issue. When flying, the spaceplane must take into consideration air regulations. We could find some indications from the navigation of sea vessels into rivers. When sailing in internal waters, sea vessels have to apply the local rules, they have to comply with both regulations. It is of course up to lawmakers to create rules, which can be applicable to both domains.

## 3.     Commercialisation of Space Transportation

### 3.1   *The Launch as the Decisive Moment for a State's Control over Space Activity*

As far as States directly controlled launching, they were in a position to control every other space activity. This is the reason why, as early as 1963, in the United Nations General Assembly Resolution, liability for damage was put on the launching State. Consequently, private launching activity is a very important change as far as space law is concerned. It is the reason why every domestic space law deals mainly with launch activity and thus with launch licensing.

### 3.2   *States' Responsibility and Liability for "National Space Activities"*

At first sight, the current system with its States' responsibility and liability may seem to impede private activity in outer space. The multiplication of launching States for one launch according to the four criteria used by the treaties and the fact that this liability is "joint and several" may be presented as complicated and hampering.

Let us have a look at these rules. The first obligation of States follows from article VI of the 1967 OST: "States are internationally responsible for any national activity carried on by governmental agencies or by non-governmental entities". This obligation goes further than the usual obligation of States related to private activities but it is merely an obligation to assure that national activities are carried out in conformity with international law and more precisely international space law. This is very useful to ensure a correct application of international law and regulations in Outer Space. For instance, when rules on the mitigation of space debris will be accepted, the obligation of States to control their implementation effectively will be strengthened.

States are also liable for possible damage. According to OST article VII, launching State(s) is (are) liable for any "damage caused by the space object". The liability Convention, article II, specifies that the launching State(s) would be absolutely liable for a damage "on the surface of the Earth or to aircraft flight" and that its fault should be proven if the damage is caused in outer space. This very much *"victim oriented"* solution was rather simple as far as only one or two States were involved in a launch. Given the alternative nature of the four criteria, the State that launches or procures the launching of an object in outer space, or the State from whose territory or facility an object is launched, many States may qualify. This is often the case nowadays. Some complications may occur. The launching States are jointly and severally liable, i.e., the victim may ask for the whole compensation from any of them. They will probably try to get compensation from the richest or from the State that is the more likely to pay.

The States have to be aware of that. It is not always the case, even for highly developed States. They should either pass agreements such as the one referred to at article V of the liability convention or make sure that, at the level of the launching contracts and insurance, the liability issue is fully addressed according to their commitments [Reference 6].

3.3   *The Burden of Risk*

The States' obligation does not necessarily constitute a heavy burden for the States. Responsibility and liability of the States, according to article VI and VII of the OST, must be distinguished on the one hand, and the issue of sharing of the financial burden considered on the other hand. Both are connected, but they are different. According to the treaty and the liability convention, the launching State would have to pay compensation to the potential victims. It does not necessarily have to carry indefinitely the burden of the compensation. It may very well obtain from private entities a refund directly or through insurance.  This refund or insurance is, most of the time, provided for in national space legislation.

When we are considering damages, which may occur in thousands of years, it is anyway difficult to find a satisfactory solution. In any case, it should be quite inadequate to leave the liability on private entities, which may become bankrupt. Some recent events show the utility of the current system on the issue of liability as well as those of responsibility and permanent jurisdiction and control.

## 4.     Normalisation of Space Transportation

### 4.1   Normalisation of Contracts

For the time being space transportation contracts are very much influenced by the hazardous nature of space activity. It can be presumed that, if space transportation changes and becomes a more "normal" activity, some consequences will follow on legal techniques. To take an example, contracts of launch use very generally waivers of liability and waivers of claims. These kinds of clause are very widely used: they are even made compulsory by the US commercial space launch act of 1984, revised in 1988.  In the case of gross negligence and wilful misconduct, it is not sure that such clauses will always be considered as valid by the judge. To take the example of French law, these clauses will certainly be very questionable in the case of gross negligence. The conjunction of waiver of liability (clause exonératoires de responsabilité) and waivers of claims (engagements de non recours) have a good chance to be considered as excessive. The commitment not to claim is generally considered reluctantly by judges.

### 4.2   Astronauts or Tourists ?

Current space law only deals with astronauts (Outer Space Treaty) or personnel of a spacecraft (Rescue agreement). If others persons are using space transportation the rules must be reconsidered. According to the Outer Space Treaty, astronauts are " envoys of mankind in outer space" but they are not considered as third parties for the liability convention that does not apply to them. Article VII excludes the application of the liability convention to nationals of the launching State and to foreign nationals participating in the operation. These rules may cause some difficulties when tourists will go into Outer Space; if safety improves, the law will have also to be improved.

### References
1.     Lafferranderie, G. (editor): *Outlook on Space Law over the Next 30 Years*, Kluwer, Devender, The Netherlands, 1997
2.     Cheng, Bin: *Studies in International Space Law*,  Clarendon Press, Oxford, 1997
3.     Dunk, F.v.d..: *Private enterprise and public interest in the European "spacescape"*, University of Leiden, 1998
4.     Kerrest, A.: Launching Spacecraft from the Sea and the Outer Space Treaty: The Sea Launch Project, *Proceedings of the IISL/IAFColloquium*, IISL -97-IISL.3.15 Turin, 1997
5.     Office of Outer Space Affairs (Vienna) : *Questionnaire on possible issues with regard to aerospace object*, COPUOS A/AC.105/635, February 15, 1996
6.     Fenema, P.v.: *The international trade in launch services*, University of Leiden, 1999

# Launch Industry Evolution and Considerations of Future Safety Regulations

**R. Gress,** Federal Aviation Administration, Office of the Associate Administrator for Commercial Space Transportation, 800 Independence Avenue, SW, Washington, D.C. 20591, USA

e-mail: ron.gress@faa.gov

### Abstract
Commercial launch operations began in the United States in 1989. Over the past 11 years, the scope of the launch industry has continued to expand in the United States. The number of regulated launch operations has increased and the development of commercial launch sites has occurred. Launch operations are now taking place from locations not operated by the US government. Proposed reusable launch vehicles and reentry operations have placed additional pressure on the need for legislation and regulations to keep pace with these changes. A summary of the recent legislative changes and regulatory initiatives of the Federal Aviation Administration are presented. Thoughts on the possible near- and long-term future regulatory issues and legal questions are offered.

## 1. Background

Since the first licensed commercial launches took place in 1989 in the United States, there have been over 125 licensed launches. In fact the number of commercial launches per year has recently exceeded those conducted by the government. Within the last four years, the Federal Aviation Administration (FAA) has licensed the operation of four commercial launch sites. Three of these sites are located on existing government launch ranges. In addition, the FAA has been challenged by companies desiring licenses to launch from non-government ranges. For most past launch operations, the FAA has been able to rely on the requirements and oversight provided by the government range operators (e.g., the United States Air Force or NASA) in considering the license application. Examples of launch operations away from government ranges include the launch by Orbital Sciences of the Pegasus vehicle from Spain and the launch operations of Sea Launch with their Zenit-3SL vehicle from a mobile platform located in the Pacific at the equator. This evolution in the scope of commercial operations drives the need to develop standards and requirements for launch operations that achieve safe operations, but without placing an unnecessary burden on the industry.

The desire to further reduce the cost of getting to orbit is also driving the development of various reusable launch vehicle (RLV) concepts by numerous entrepreneurial companies in the United States. Such concepts raise the potential for a whole new spectrum of additional uses in the commercial

201

*M. Rycroft (ed.), The Space Transportation Market: Evolution or Revolution?*, 201–208.

environment including the carriage of passengers, recovery and repair of spacecraft, launches and landings of the same vehicle occurring at different locations around the world, and space station servicing. Dramatic reductions in launch costs also increase the likelihood of the development of other commercial uses of space beyond communications and remote sensing. RLV concepts also present extensive immediate and long-term safety issues and the corresponding need for the development of safety standards and regulations.

## 2.    FAA Authority and Responsibilities

### 2.1    Original Authority and Responsibilities

In 1984 Congress passed legislation stating that no one may conduct a launch or operate a launch site within the United States without a license from the Secretary of Transportation [Reference 1]. This authority is delegated to the Associate Administrator for Commercial Space Transportation in the FAA. The scope of FAA's responsibility under this legislation includes the responsibility for:

- Public health and safety

- Safety of property

- National security

- Foreign policy interests of the United States

- Insurance requirements (Government property and third party liability claims).

Any United States citizen conducting such operations anywhere in the world would also require a license.

In addition to licensing, the FAA may also:

- Conduct safety compliance inspections

- Impose enforcement actions

- Investigate accidents.

## 2.2   Recent Changes to FAA's Authority

In 1998 Congress provided additional authority to the FAA that added requirements for the licensing of reentry operations and reentry landing sites [Reference 2]. This provided a parallel to the requirements for launch and launch sites. In addition, Congress provided the authority to the FAA to approve:

- Launch and reentry vehicles

- Safety systems

- Safety processes

- Safety services

- Safety personnel.

## 3.   Current FAA Regulatory Initiatives

The FAA has recently updated its regulations originally issued in 1988 covering launch activity of Expendable Launch Vehicles (ELVs) [References 3, 4]. These regulations focus primarily on the requirements for launches from government ranges. The FAA has also recently issued regulations codifying the requirements and standards for financial responsibility of FAA launch licensees, including insurance [Reference 5].

Because of the development in the industry discussed earlier, the FAA published, and sought comment on, proposed regulatory requirements for:

- Licensing the operation of a launch site [Reference 6]

- Licensing RLV and reentry operations [Reference 7]

- Financial responsibility requirements for reentry operations [Reference 8].

In issuing proposed requirements on RLVs and reentry operations, the FAA met a congressionally mandated requirement to issue proposed rules in this area within 6 months of the statute granting FAA licensing authority. The FAA expects to issue final rules in all these areas shortly. The FAA also expects to issue a notice of proposed rulemaking covering launch activities of ELVs from non-federal launch ranges in the near future. In this area, the FAA has been working closely with the government ranges with a goal of developing a common set of core safety standards and requirements for all launch activities. Until these rules become final, the license applications have to be evaluated on a case-by-case basis.

## 4.    Future Regulatory Considerations

### 4.1    General Strategy

Clearly, the emerging technologies are the drivers of the regulatory needs in the future. The FAA has approached its development of standards and requirements for RLVs, for example, in its initial proposed rules for the launch and reentry of RLVs and other reentry vehicles at a top level, using performance requirements and standards for safety processes wherever possible. This performance-based approach provides for the greatest flexibility for innovation and certainly is critically important because of the large variation in RLV concepts being proposed. These variations in the RLV concepts being proposed are possibly greater than the variations between fixed wing, rotary wing and lighter-than-air aircraft in the aviation realm. They make it impractical to expect that the initial RLV regulations would be as detailed, and in some cases, as specialized, as those addressing the aviation realm. In addition, the FAA wishes to develop regulations that do not pick winners and losers among technological approaches within the industry, but rather achieve safe operations.

### 4.2    Future Safety Issues

While the FAA has issued a notice of proposed rulemaking covering the safety of launch and reentry of RLVs and other reentry vehicles, it has only just begun dealing with the future safety issues associated with such vehicles and operations.

**Operations and maintenance.** RLVs, by their nature of reusability, present new safety issues that are not typically associated with the operation of ELVs. Requirements for maintenance and inspection programs during the operational life of a vehicle or vehicle type must be addressed. While aircraft operations provide a model for dealing with such issues, it remains unclear how applicable such a model will be in general. Little is known regarding the effects which repeated flights might have on the functioning and reliability of key systems. The Space Shuttle provides some insights, but that is only one data point and from a vehicle that is based largely on old technology. What time periods may be appropriate between inspections and maintenance for the critical systems? A lot may, again, depend on the design of the specific vehicle and the operational envelope of the vehicle or stage.

**Crew and passengers.** One cannot talk about RLVs very long before the subject of the carriage of crew, and even passengers, comes up. Current safety standards for ELVs have focused on the safety of the public. Even the current proposed rules covering RLVs has the safety of the uninvolved public as its focus. What additional safety standards might be appropriate for a vehicle requiring a flight crew? How should initial operations with flight crews compare to aviation and its treatment of the safety of test pilots? What medical requirements or standards should be associated with the crew? In the near term, what vehicle safety and reliability requirements would be appropriate for the operation of a vehicle carrying passengers "for hire"? How should these issues be addressed and what considerations should be given to the accepted levels of safety in the aviation industry and the adventure tourism industry. Is it possible that, in the future, it might be appropriate to expand the focus of the regulatory program to address minimum standards on the success of ELV and RLV missions, not just the public safety aspects of the operation? As the industry moves into the phase of its development to include transporting people, there may no longer be any distinction between what is needed for safety and mission assurance.

**Integration into the national air space.** The FAA is working on how the operations of RLVs may best be integrated into the National Air Space (NAS) used largely by the aviation industry. The FAA currently performs these functions for the government operated launch sites. However, are there better ways of performing these functions as the number and location of launch sites increase or for RLVs? Aside from the Space Shuttle, there is not a lot of experience with reentry issues. Clearly, the characteristics of the vehicle and the location of the launch site or reentry site will play an important role in how this may best be accomplished in a particular case. With the operation of new launch sites to support ELVs and the expected growth of launch sites to support RLVs, this is a very important safety consideration. In some instances, the launch and reentry operations present unique safety issues (e.g., different flight trajectories and large differences in speeds) with respect to the safety of the operation and of aircraft operating in the NAS. Other areas being considered are whether any basic equipment requirements for the vehicle and ground systems might be necessary, and what procedures and planning might be minimally necessary for coordinating reentry operations.

*4.3   Possible Role of Approvals*

The recently granted authority to issue approvals will provide a foundation upon which other future areas likely to be of value to the industry and the FAA can be addressed. Of course requirements, standards, compliance monitoring programs and associated resource needs will have to be developed for each individual element.

**Vehicle approvals.** In addition to issuing a license to authorize an entity to conduct launch operations, the authority exists to issue an approval of the vehicle. One of the first steps in implementing this authority is carefully to determine what the issuance of such an approval means. Should it parallel the concept of airworthiness certificates for aircraft and, if so, should it imply that the same safety standards or levels are achieved?

**Vehicle safety system approvals.** Vehicle systems may be approved as meeting certain minimum performance requirements. In some cases, systems such as propulsion, attitude control, navigation, and health monitoring, to name a few, may be offered for use on different vehicles. Not unlike the situation in the aviation arena, care must still be given to how these systems perform in the integrated system of a specific launch vehicle.

**Safety personnel and process approvals.** The FAA may approve personnel performing safety-critical launch or reentry functions. Those safety functions could include company inspectors, pilots, and licensee safety officials, but work would need to be done to determine which functions merit the need for such approvals. The minimum qualification requirements would have to be developed and requirements to demonstrate continued proficiency determined. Potential processes approved could include test procedures and pass/fail criteria for certain types of ground system and integrated vehicle tests. Another example of the type of approvals which the FAA may issue include independent verification and validation of safety critical software. Inspection and maintenance procedures may also be candidates for such approvals.

**Safety services.** Here we are most likely talking about another party offering safety services to a potential licensee. The service area potentially could overlap, in many respects, the approval areas involving the vehicle, vehicle systems, and personnel and processes. The scope of approvals in the service area will depend heavily on how the industry continues to evolve. It is not difficult to envisage companies offering range safety services for launch operations from new launch sites. Other areas with parallels to the aviation model may eventually develop in the more distant future, including maintenance and repair centers.

The above examples just scratch the surface of the areas that may be of interest in the safety area. What priorities may be placed in these areas is still unclear and must be determined. Unfortunately, developments in these areas will not be easy. In looking at the aviation model, many of these approaches evolved over a long period of time and are based on nearly a century of experience in aviation. Most commercial aircraft are mass-produced and collectively provide many more flight hours of experience than could be achieved in the flights of the first RLVs. It appears that, in some areas, it may take several years for the industry and the FAA to develop the experience, and to collect the data, needed to support initiatives.

## 5.    Potential International Considerations

Other areas that may need to be investigated at some point in the future include the need to address international issues. Much like other global transportation systems such as aviation and marine transportation, space transportation may become an industry where company operations routinely cross international boarders. We can already see the start of the international character of launch services with International Launch Services, Orbital Sciences and Sea Launch. With RLVs, one can envisage reusable launch vehicles launching and landing in various countries as the transportation demand dictates. Are the international treaties such as the Treaty on Principles Governing the Activities of States in the Exploration and Use of Outer Space, Including the Moon and Other Celestial Bodies, of January 1976, and the Convention of International Liability for Damage Caused by Space Objects of November 29, 1971, adequate to address these types of operations and activities that may occur in the future?

The United States is one of the few space faring countries that currently has in place regulations covering the operations of the commercial launch industry. Where other countries have begun to develop their own laws covering the safety of such activities, the approach of the United States has served, to some extent, as a model. It is in the interests of the worldwide launch industry to continue to operate in a safe manner. In this global economy, everyone is affected when a launch operation fails. It is almost inevitable that as new vehicles and concepts are introduced, they will suffer setbacks. It is imperative that safety be maintained, a responsibility that rests with all counties involved in launch operations.

**References**
1.  49 United States Code, Subtitle IX, chapter 701, Commercial Space Launch Transportation Activities, as amended
2.  HR 1702, Commercial Space Act of 1998
3.  14 CFR, Chapter III, Parts 400-415, Commercial Space Transportation Licensing Regulations, 1988
4.  14 CFR, Chapter III, Parts, 401, 411, 413, 415, and 417, Commercial Space Transportation Licensing Regulations, 1999
5.  14 CFR, Chapter III, Part 440, Commercial Space Transportation Financial Responsibility Requirements for Licensed Launch Activities, 1998
6.  Federal Register Notice, *Licensing and Safety Requirements for Operation of a Launch Site*, Notice of Proposed Rulemaking, June 25, 1999
7.  Federal Register Notice, *Commercial Space Transportation Reusable Launch Vehicle and Reentry Licensing*, Notice of Proposed Rulemaking, April 21, 1999
8.  Federal Register Notice, *Financial Responsibility Requirements for Licensed Reentry Activities*, Notice of Proposed Rulemaking, October 6, 1999

# Establishing a Space Launch Industry:
# The Political and Regulatory Considerations

**M. Davis**, Ward & Partners Lawyers, Australia

e-mail: mdavis@wardpartners.com.au

**Abstract**

The Australian space community has recently experienced a period of optimism as a result of the announcement of several commercial launch projects. The Australian Government has responded by establishing a space launch licensing regime through legislation. However problems with funding have highlighted the practical problems associated with the creation of a new commercial launch industry in a very competitive international environment.

Drawing mainly on the experience of the United States, this paper examines the factors that are necessary for a sustainable commercial launch industry and suggests that the role of governments is critical. While an appropriate regulatory regime is a good first step, a commercial space industry in Australia needs strong political support and pro-active government policies to share risk and enhance the viability of the industry.

## 1.    Introduction

"The emergence of a commercial space transportation industry is, in many ways, reminiscent of other forms of transportation – maritime, rail and air. All presented technological challenges and risks, all initially involved periods of trial and error and all ultimately assisted in fostering economic growth and enhancing the quality of life", [Reference 1].

The Australian Space Activities Act was enacted in December 1998, at a time of optimism in the Australian space community. Australia's defunct Woomera spaceport had been chosen as the site for the testing of the Kistler K-1 reusable launch vehicle. Work on the project in Australia was about to start and the government was expressing strong support for the establishment of a new commercial launch industry. In the government's view, the need to legislate for a regulatory regime was pressing [Reference 2].

When introducing the Space Activities Act in 1998, the government placed emphasis on setting out the legislative framework necessary to facilitate commercial space launches from Australia. To quote Senator Kemp [Reference 2] again, "A clear legislative and regulatory framework is essential for any of these projects to succeed."

The new legislation resembled the US Commercial Space Launch Act [Reference 3]. A new bureaucracy sprang up in Canberra to administer the

*M. Rycroft (ed.), The Space Transportation Market: Evolution or Revolution?*, 209–217.

licensing and safety provisions of the Act, to formulate policy and to develop the detailed regulations necessary for the complex process of assessing the risk to the public and evaluating the technical and financial aspects of launch projects. This new bureaucracy has been kept busy working with three other Australian proponents of commercial launch projects that have emerged in the past three years.[1]

Eighteen months down the track, the optimism has been replaced by sober pronouncements by some that, yet again, the Australian commercial launch industry may be stillborn [Reference 4]. A shift in the international investment climate has prevented Kistler from obtaining the remaining funding needed to proceed with its test program. Given the delay and the changing shape of the market for low Earth orbit launches, there must be some question as to its readiness to fulfil its original launch orders. The other proponents face similar financing hurdles and are yet to announce the commencement of work on infrastructure facilities.

Australia's space heritage dates back to its prominent role in the space race of the Cold War. Through the provision of launch facilities at Woomera, the Australian Government became an important participant in an impressive international satellite tracking and space launch program. It also developed expertise in satellite engineering. At its peak, Woomera was one of the largest spaceports in the world [Reference 5]. The indigenous technical capability in satellite engineering and space science that Australia developed during this

---

[1] Asia-Pacific Space Centre Pty Ltd proposes to establish a launch facility on Christmas Island, an Australian territory to the North West of Western Australia. The company plans to use Soyuz, or possibly the new Russian Angara type rockets, to launch satellites into low Earth orbit and geostationary orbit. As a prerequisite for the obtaining of the necessary space licence, an environmental impact study has recently been prepared. For further information, see <http://www.apsc.com.au/>.

United Launch Systems International Pty Ltd is an Australian company with a major investment by the Thai Satellite Telecommunications Company proposes to establish a low Earth orbit space launch facility using the newly developed Russian built Unity launch vehicle. The proposed launch site is at Hummock Hill Island, 60 km South of Gladstone on the eastern coast of Australia. Test launches are anticipated in 2002 with commercial operations commencing in 2003. For further information, see <http://www.ulsi.com.au/>.

Spacelift Australia Limited, an Australian company, announced in September 1999 that it had entered an agreement with a Russian company, STC Complex – MIHT, to establish a low Earth orbit launch service from Woomera commencing in late 2000 or early 2001. The agreement involves the use of the Russian Start launch vehicle. For more information see <http://www.spacelift.com.au/>.

period has survived to this day and, despite relatively low funding, Australians are involved in a range of national and international projects.

## 2.    Why Launch from Australia?

Australia is cited as an ideal country of launch because of its location, its extensive land areas of low population, its well developed and sophisticated infrastructure, including telecommunications and transport facilities, its stable political history, its strategic alliances with western powers, its respected non-proliferation credentials and its space heritage.

However, these undeniable advantages are not, of themselves, sufficient catalysts for the creation of an economically viable launch industry. The hiatus between the space activity of the 1960's and 1970's and the commercial launch projects of today has not helped. During the last two decades, the objective benefits of establishing a commercial launch industry were identified and projects were proposed. However, Australia lacked two fundamental drivers — a political commitment to a properly funded national space program and the political will to assist to establish Australia as a commercially competitive country of launch [Reference 6].

The lack of a properly funded national space program over the past two decades cannot be underestimated. As Mr Edward Frankle, NASA's General Counsel, said recently, "It is now clear to all spacefaring nations that any successful national space program must rest upon the twin pillars of governmental and commercial activities [Reference 7]."

Given this scenario, is it feasible for Australia to develop a commercial launch industry? The theme of this paper is that success is only possible if there is a convergence of political and economic forces in the near future. In spite of the best endeavours of those involved in a series of past government and private initiatives, success has not been achieved because the right combination of market conditions and government policy has been missing. The willingness of the government to establish a domestic regulatory regime and the recent efforts of the Australian Department of Industry, Science and Resources to smooth the regulatory path and to develop a coherent space industry policy are encouraging signs, but much more needs to be done.

Internationally, the emergence of a commercial launch industry is a relatively new phenomenon. The most successful commercial launch company has been Arianespace. Ironically, its origins derive from the launches that were conducted from Australia.[2]

## 3.     The Role of Governments

What are the influences that have allowed commercial launch ventures in other places to succeed? Is it possible to learn lessons from the experience of other countries? Reviewing the short history of the global commercial space launch industry, the main lessons derive directly from the role of governments, both in a preventative and in a pro-active sense.

### 3.1     Governments must Assist by Sharing or Reducing the Many Forms of Risk that Private Ventures Face

Commercial launch companies face many varieties of risk, on a scale that is not comparable to most other industries. Apart from the significant capital costs required for launch infrastructure and ground services, safety and environmental Considerations require that exhaustive studies and surveys are performed and that elaborate preventative actions are implemented. These add considerable cost and time delays to the projects.

Because the initial capital costs are so great, start up companies with commercial launch plans can encounter considerable resistance from potential investors and cannot expect to obtain all of their funding from private investors or financial markets.

Another risk relates to the market perception of an inexperienced launch provider. While launch prices are undoubtedly an important factor for launch customers, other considerations in the choice of launcher include the track

---

[2] The European Launcher Development Organisation (ELDO) came into being on 29 February 1964.   The members of ELDO were Belgium, The Federal Republic of Germany, France, Great Britain, Italy and the Netherlands.  Australia was a full member contributing its facilities in lieu of a financial contribution.   The launch vehicle developed by ELDO was known as Europa.  It consisted of a modified Blue Streak provided by Britain as the first stage; the second stage, called Coralie, was provided by France and West Germany provided the third stage, known as Asteris.  The test satellite was developed with the leadership of Italy, while the Netherlands and Belgium were responsible for the development of telemetry and guidance systems. Between June 1964 and June 1970 there were ten test firings of the Europa rocket from Woomera. In 1968 Britain decided to withdraw from ELDO for economic and political reasons. Because of the desirability of launching communication satellites into geo-stationary orbit from near the equator, it was decided to pursue further launches from France's equatorial launch site at Kourou in French Guiana.

record of the launch company, the availability and reliability of the launcher, the cost and availability of insurance, and the quality of facilities and services at the launch site.

Governments can assist in a number of ways. First, they can provide financial assistance. In addition to direct funding, it should be possible to provide ground facilities and infrastructure at reduced cost to the launch provider. Secondly, governments are needed to apply diplomatic pressure to facilitate inter-governmental agreements necessary for the export of payloads and, in some cases, for the international transfer of launch technology expertise. Thirdly, governments can promote confidence by assisting in the international marketing of launch services, particularly to other governments. Many countries still regard space activity as the natural domain of governments, and are wary of private ventures offering launch services that do not have close associations with governments.

### 3.2 Governments should Play a Direct Role in Minimising Exposure to Legal Liability for Damage

One of the main risks associated with orbital or sub-orbital launches is the risk of damage to humans or property as the result of a catastrophic accident. This is a risk management issue that must be carefully addressed in any commercial business plan.

The risk of exposure to unlimited liability for compensation for accidental damage was such an issue for US companies in the early 1980's that legislative intervention was necessary [Reference 1]. The US Congress responded by agreeing to relieve launch operators of that unlimited exposure:

"The Government must act as a reliable partner and be willing to provide transitional support to this nascent industry until such time as equitable rules of the road are in place and the insurance industry has the capability to accommodate the increased insurance requirements of commercial space activities at reasonable rates" [Reference 8].

As originally enacted, the US Commercial Space Launch Act (CSLA) required a licensee to have (in effect) liability insurance, at least in such amount as the government deemed necessary having regard to the international obligations of the United States. Under the amendment in 1988, the obligation to obtain insurance or demonstrate financial responsibility was limited to the maximum probable loss from claims arising from the launch, with a maximum insurance level of $ 500 million or "the maximum liability available on the world market at reasonable cost" [Reference 9].

By adopting a regulatory regime modelled on the United States system, the Australian government has agreed to shoulder a similar risk.[3] While it can be argued that the acceptance of unlimited international liability for damages claims above the Maximum Probable Loss is in itself a considerable assistance to private industry, only one international claim has ever been made.[4] Therefore it can be stated that while the financial exposure is high the probability of a financial payout by a government is low.

3.3  *Governments should take an Active Role in Promoting and Assisting the Establishment of a New Space Launch Industry through Deliberate Pro-industry Policies*

In Europe and in countries such as Russia, Japan, China and India, governments have directly funded launcher development. Determining the financial viability of the commercial launch services which they now provide is somewhat academic. However, the United States has an active private launch services industry, and this is the industry model from which the main lessons for Australia can be learned.

In the early 1980's the US Government began to examine ways to promote the domestic launch industry. Among the factors that were generally regarded as inhibiting the establishment of such an industry were:

- The stranglehold on the market for commercial satellite launches by NASA (the Space Shuttle) and Arianespace

- The risk of exposure to unlimited liability for catastrophic launch accidents

- The lack of a single agency responsible for authorisation and continuing supervision of 'national activities in outer space' as required by Article 6 of the Outer Space Treaty [Reference 10].

---

[3] One further benefit of the Australian legislation is that, unlike under the US legislation, there is no ceiling to the government's liability above the maximum probable loss. Section 16(a)(1)(a) of the CLSA, which means that the licensee must assume liability for losses in excess of $1.5 billion.

[4] This was a claim by Canada against the Soviet Union following the uncontrolled re-entry of COSMOS 954 in the Northwest Territories of Canada on 24 January 1978. It was settled for the modest figure of $3 million (Canadian). See Van C. Ernest, *Third Party Liability of the Private Space Industry: To Pay What No One Has Paid Before,* (1991) 41 Case W. Res. 503.

The US Commercial Space Launch Act of 1984 [Reference 9] established a launch licensing regime but went beyond the mere regulation of the industry. Its expressed purposes (see Section 3 of Reference 9) were:

- Promotion of economic growth and entrepreneurial activity through utilization of the space environment for peaceful purposes

- Encouragement of the United States private sector to provide launch vehicles and associated launch services by simplifying and expediting the issuance and transfer of commercial launch licenses and the use of government-developed space technology, and

- Designation of an executive department to oversee and coordinate the conduct of commercial launch operations, to issue launch licenses and to protect the public health and safety, safety of property and national security and foreign policy interests of the United States.

The development of the US commercial launch industry was therefore promoted by direct government action, making government facilities available on a cost reimbursable basis and promoting the use of commercial launch vehicles and services by US Government agencies.[5] There was a further policy change in August 1996 when President Reagan, following the Challenger tragedy, decided to prohibit commercial satellite launches from the Shuttle. However, the US domestic launch market did not really take off until the US Congress in 1990 directed NASA to purchase launch services from commercial providers for the majority of its launches [Reference 7].

It can be seen that a strong political commitment to assisting the development of a private launch industry in the United States was essential. Policies that demonstrate the government's active support, both domestically and internationally, for the industry need to be established. This goes well beyond creating a regulatory system and declaring a level playing field for the private players. It includes the type of support that the US commercial space launch industry has received and continues to receive from its government.

---

[5] The policy seeks to encourage private development of outer space by "'bring[ing] into play America's greatest asset -- the vitality of our free enterprise system.'" *Comment, Legal Aspects of the Commercialization of Space Transportation Systems*, 3 High Tech. L.J. 99, 102 n.19 (1989) (quoting President's Radio Address to the Nation, 20 Weekly Comp. Pres. Doc. 113-14 (Jan. 28, 1984)).

## 4.     Conclusion

The international lessons are clear. For the creation of a viable domestic launch industry, it is a mistake to assume that all that is necessary is a competitive commercial environment and a regulatory regime that complies with international legal obligations. To quote Frankle [Reference 7] again, '... past experience indicates that the behavior of industry and government is motivated more by national commercial and economic self-interest than by international legal requirements.'

In the Australian context, the government lacks the ability to assist the launch industry by directing its own business to commercial operators, but opportunities will arise for the government to exercise influence over the choice of launcher from time to time. Because most of the commercial launch business will come from foreign customers, government policy should be directed to innovative approaches to promoting Australian launch services as an export industry. [6] This would include:

- Ensuring that the development of the industry is given national priority and, similar to the commitment made in the 1980's in the United States and in France, creating a partnership between government and the private sector in which the risk is shared. The US Government and Congress took an activist role in promoting a commercial space launch industry by developing policies to ensure that the fledgling industry became viable. This contrasts with the relatively passive role of the Australian government to date

- Providing financial assistance, financial incentives to launch customers, tax breaks and relief from government charges, and free or subsidised access to government services and facilities

- Ensuring that the regulatory regime for launch operators and their customers and also the practical administration of laws is simpler and less onerous than in other countries

- Working with launch providers to ensure that the regulations concerning third party launch insurance requirements in particular and risk assessment generally can be easily understood and calculated, that the

---

[6] Australia has a liberal telecommunications licensing regime that allows foreign companies, including satellite operators, to obtain carrier and frequency licences. These companies are required to develop business plans that will benefit Australian industry.

insurance is readily available, and that insurance premiums and other risk prevention requirements do not amount to a cost disadvantage

- Using international influence to smooth the way for cooperation between governments and government agencies (in particular in the United States) to facilitate the restrictions on the export of satellite technology for launch from Australia.

The challenge which Australia faces is that it is entering a highly competitive international industry without the advantage enjoyed by larger countries of a government assisted space industry from which opportunities for commercial operators can be leveraged. There are powerful reasons for choosing Australia as a country of launch, and if the policy settings and international agreements are in place, there should be no difficulty in ensuring that a sufficient share of the international launch market comes our way. Measures such as those outlined above are needed to ensure that launch operators in Australia are able to offer competitive launch services. An industry that will provide a sustainable competitive advantage for Australia, and that will benefit the economy of the country as a whole, will then be possible.

### References
1. Nesgos, P., *Commercial Space Transportation: A New Industry Emerges*, (1991) Vol XVI Annals of Air and Space Law 393
2. Second Reading Speech by Senator Kemp, *Australian Senate Hansard*, 12 November 1998, p. 148
3. Commercial Space Launch Act (US), Pub.L. 98-575, 49 U.S.C. 2601 (30 October 1984) as amended by The Commercial Space Launch Act Amendments of 1988, Pub.L. 100-657
4. Middleton, B., *The Global Space Launch Market in the Next Decade*, Paper delivered to 12th National Space Engineering Symposium, Adelaide, September 1999.
5. Dougherty, K. and James, M., *Space Australia: The Story of Australia's Involvement on Space*
6. Griffy Brown, C., Knapp, K. and Humphreys, J., *The Cape York Spaceport: Lessons in Sustainable Strategic Advantage for Developing Commercial Space Ventures*, <http://www.gbhap-us.com/fulltext/free/S960040F793.htm>. The Gordon and Breach Publishing Group, Space Forum, 1996, (accessed 21 April 2000)
7. Frankle, E., *Legal Considerations Affecting Commercial Space Launches From International Territory*, Paper delivered to IISL Symposium, Amsterdam, October 1999
8. *Report of the Senate Committee on Commerce Science and Transportation on H.R. 4399*, 100th Congress, 2nd Session, Report 100-593, Washington 1988 at 3
9. Commercial Space Launch Act, Section 16(a)(1)(a)
10. (1967) 610 United Nations Treaty.Series 205

# Internationalisation of RLV Regulations:
# A Realistic Future Need

**C. Jolly,** Consultant, Paris, 75013, France

e-mail: claire.jolly@wanadoo.fr

Abstract

Considered a few years ago as only futuristic projects, Reusable Launch Vehicles (RLVs) have become in the United States, and elsewhere in the world, the subjects of important national and private investments in research and development. At the same time, governmental agencies, especially in the United States, are taking the first steps towards an innovative RLV regulatory system. As prototypes will give way in the next decades to a new generation of potentially diverse launch vehicles, new international policies and regulations will have to be created to give a common framework to this growing and diverse industry.

## 1. RLV Developments Everywhere

Developments of space transportation reusable technologies have become in the last five years a priority for many space industries worldwide. American governmental agencies are already at the stage of developing demonstrators with industry (see Table 1), while companies in Europe and Japan are also investing in research and development for reusable launch vehicles (RLV) technologies (see Table 2). Even though RLV developments might take longer in Europe and the rest of the world than in the United States, their research programs might result in operational RLVs that need to be taken into account in future space launch regulatory systems.

| Companies | RLV | Description / Comments [1] |
|---|---|---|
| Lockheed Martin | VentureStar | - SSTO – start of operations 2012-2015<br>- Follow on to X33 demonstrator, decision to develop or not in 2001 |
| Kistler Aerospace | K-1 | - TSTO – start of operations in 2001-02<br>- Private initiative facing financial rather than technical problems |
| Rotary Rocket | Roton | - SSTO – start of operations 2002-2003<br>- Technical feasibility still under debate |
| USAF | Spacelane | - Under development with US industry<br>- Operations in ~2010 |

**Table 1.** Overview of some American efforts towards RLVs

---

[1] SSTO: Single-stage-to-orbit, TSTO: Two-stages-to-orbit.

*M. Rycroft (ed.), The Space Transportation Market: Evolution or Revolution?*, 219–226.
© 2000 *Kluwer Academic Publishers. Printed in the Netherlands.*

In May 1999, the European Space Agency (ESA) Council at ministerial level approved the FLTP (Future Launch Technologies Program), which is an optional program investigating concepts and technology requirements for future reusable launch vehicles or semi-reusable launch vehicles. Seven ESA national delegations have committed themselves to participate in the FLTP (France, Belgium, Netherlands, Spain, Austria, Sweden, and Switzerland). Germany is pursuing a national program of its own but might participate in a later phase of the European program.

| Companies Europe : | RLV Developments | Description / Comments |
|---|---|---|
| Aerospatiale Matra | Taranis | - Three step program to develop a RLV (ESA's Future Launcher Technology Programme) towards 2015-2020<br>- 2 experimental aircraft ARES (suborbital and transonic air-dropped demonstrators) and THEMIS (cryogenic-powered demonstrator) |
| Dassault | Vehra demonstrator | - suborbital demonstrator<br>- might be integrated in Aerospatiale's RLV program |
| Dasa | Hopper | - TSTO – potential operations in 2012<br>- Follow on to Phoenix demonstrator |
| Japan: | | |
| Mitsubishi, Kawasaki | Hope X demonstrator | - Objective is to develop a SSTO<br>- Operations in 2015-2020 |

**Table 2.** Some worldwide efforts towards RLVs

Other actors in Asia are investing in RLV technology. Japan is exploring numerous technologies (i.e. air breathing propulsion) through specialized development programs. China, in its endeavor to make human spaceflight missions, is developing recoverable Soyuz-type capsule and has plans to develop a space plane in the future, while India has also declared an interest in reusable space technology [Reference 1].

The financial investments and political will of all those countries to develop reusable technologies might not be as strong as in the United States but, as RLV developments are becoming a reality throughout the world, most analysts expect reliable new space transportation systems to be operational in 2012-2020. New technology advances and impacts cannot be ignored and, therefore, many international regulations will need to be adapted or even created.

## 2.    Regulatory Impacts of RLV Missions and Flight Profiles

Space launch activities are regulated by national space laws (i.e., national licensing, regulations of spaceport activities) and by international treaties (i.e., registration of space objects, international liability of State of registry). But the new wave of space transportation activities, which comprises new RLV technologies and an increasing number of public-private partnerships will bring new sets of regulatory issues that have to be addressed [Reference 2].

### 2.1   New Space Transportation Means... New Regulations?

RLVs' flight profile and missions will have impacts worldwide. International space law, national policies and air-traffic regulations will need to evolve.

If we consider the future activities of RLVs, such as transportation between the Earth and orbits around the Earth, and transportation between places on Earth via outer space, different national and international liability issues will have to be considered. These are currently under study, in the United States in particular [Reference 3]. Cargo and crew transfer to and from the International Space Station are examples of future potential RLV missions, as are fast package-delivery or 'space tourism' sub-orbital flights. Therefore, a domain that needs to be carefully regulated will be the future aerospace traffic system. Reusable launch vehicles will transit to and from space using the same airspace that is currently the exclusive domain of aviation traffic.

There has already been a debate, for many years in the United States, on whether new RLV regulations should be largely inspired by Aviation Law or by Space Law [Reference 4]. Indeed, two main views pertaining to the enactment of RLV regulations generally prevail. Either RLV specific regulations should be created, or current aviation regulations should be applied to space vehicles.

The current policy in the United States is a mix of the two types of regulations. But the diversity of the current RLVs under development does not facilitate the enactment of a global system. The different flight profiles under consideration are so diverse (i.e., vertical take-off/horizontal landing, horizontal take-off/horizontal landing, air dropped/horizontal landing...), that those vehicles will bring different kinds of liability implications, not only towards their own payload and passengers, but also towards third parties.

## 2.2   Current Status of 'RLV Space Policies' in Different Countries

Many countries, including the United States and Australia, are developing policies and regulations to accompany the developments of those new space transportation vehicles.

The United States is at the forefront in this aspect. A flexible licensing regime has been approved by the Federal Aviation Administration (FAA) to allow the American regulations to develop along with the RLVs' developments. The different RLV systems' configurations under development will be evaluated individually by the FAA. Negotiations with the FAA should follow in order to file an agreed Licensing Plan. The process is, for now, very individualistic and its efficiency has yet to be proven, each company being able to interact and negotiate with the FAA on a bilateral basis [Reference 5].

It is anyhow a step in the right direction. Before the Commercial Space Act of 1998, only the launching aspect of private space vehicles was regulated. This is because expendable launchers' stages are destroyed a short time after taking off. With the increasing number of private RLV companies, new procedures to regulate the reentry and landing of reusable vehicles suddenly seemed important to establish. That need for regulations prompted a fear in private industry that a too restrictive regime would "either bind the creative aspects of a company's particular RLV design, or delay a project and put the backing company out of business" [Reference 6]. The continuing cooperation between industry and governmental agencies in establishing new regulations, as hot-tempered as it might sometimes be, is a good source of inspiration for other countries.

Indeed, other governments have in the last three years developed new space policies, taking into account some RLV specificities. Interesting developments concerning space launch licensing have occurred in Australia and the United Kingdom. Seeing an increasing number of private American launch companies interested in their launch facilities, the two governments have created new space policies in order to accommodate the new RLV firms [2].

In Europe, moves are slowly being made to create a common space policy framework. As European industry is actively engaging in reusable technology studies and hopefully building demonstrators in the short-term (i.e., the ARD capsule demonstrating reentry technology), RLV policies and regulations will have to be drafted. Therefore, even if Ariane V is to be the main European launcher for a while still, the current plans to make some of its components

---

[2] Kistler is building its first spaceport in Woomera, Australia.

reusable and to work on its successor – maybe a RLV– will bring forward the need for new regulations.

It is clear that RLVs are the subject of many discussions worldwide. Technological developments are taking place and national space policies are slowly being adapted. But, as for Aviation Law or the Law of the Sea, sooner or later, an international framework for aerospace operations will need to be created.

## 3.    The Need of an International Framework

The current trend towards reusable space technologies has recently become a recurring subject of debate in the United Nations and other national and international organizations.

### 3.1    *Discussions in the United Nations Affecting RLV Regulations*

Since its establishment by the United Nations General Assembly in 1959, the Committee on the Peaceful Uses of Outer Space (COPUOS) has been the focal point of international political and legal discussions regarding outer space issues.

Public-private partnerships are still at the core of space launch activities, as governments have always been in control of research and development efforts. But the current trend towards a more commercial use of space could bring in the coming years a more active role of private actors, especially in the space transportation arena. Under international law, States are still internationally responsible even if space activities are carried out by their private entities (Article VI of the Outer Space Treaty).

Therefore, following the spur of new space activities, many international space law issues, including the concept of the launching State and the question of the "boundary" between air space and outer space (i.e., freedom of outer space versus the sovereignty of air space), are currently under discussion at the United Nations.[3]

All those issues will probably bring in the coming years a revised international legal framework, that will be accompanied by national RLV laws

---

[3] The Legal Subcommittee of the Committee on the Peaceful Uses of Outer Space had its thirty-ninth session in Vienna from March 27 to April 7, 2000, and started to review the concept of the "launching State".

and policies (i.e., national licensing). But other steps must be taken to start elaborating an international RLV regulatory system.

### 3.2  Regulating RLVs Internationally: A Special Role for the International Civil Aviation Organization?

There are already global structures for cooperation concerning air transport like the International Civil Aviation Organization (ICAO), the International Air Transport Association (IATA) and the Airports Council International (ACI). But no global structure exists yet for space transportation activities.

Regulations governing international air transport are contained in the Convention on International Civil Aviation and its 18 Annexes. They are referred to as Standards and Recommended Practices (SARPs) and deal with such issues as personnel licensing, aircraft operations and airworthiness of aircraft. Under this Convention, responsibility for implementing the SARPs rests with the 185 Members States of ICAO. But an Universal Safety Oversight Audit Program is regularly conducted by the organization and consists of regular, mandatory, systematic and harmonized safety audits performed in all of the Member States. This audit assesses the degree to which the countries have implemented the SARPs with a view to identifying and correcting deficiencies and shortcomings in safety-related areas.

Because of its international influence and its expertise on international aviation issues, the ICAO might be a good organization to start drafting common international rules for future RLV operations. Already in the late 1970s, Menter had suggested that the ICAO be made competent to regulate some of the space flight profiles of future space vehicles, so that the day-to-day international air-traffic would not be interrupted [Reference 7].

Furthermore, the ICAO is currently working on the worldwide implementation of "CNS/ATM systems", which involves the application of satellite and computer technologies to Communications, Navigation, Surveillance and Air Traffic Management. This global intertwining of national and regional systems will have to include room for future RLV operations.

Dr Kotaite, ICAO Council President, stated recently that ICAO was the "logical international institution to lead the way into space, and should work with its member States and other international organizations to provide the guidance on space management" [Reference 8]. But there might be another way to regulate and coordinate future space transportation operations.

In many parts of the world civil aviation agencies are working on upgrading their national and regional air-traffic systems. In the United States, the Federal Aviation Administration is working on its *Concept of Operations for Commercial Space Transportation in the National Air Space in 2005.*

This document has been developed by the Associate Administrator for Commercial Space Transportation (AST) in anticipation of the evolution of the future American aerospace environment, which will need to integrate an increasing air-traffic with commercial space operations. The creation of special "Space Transition Corridors" is already anticipated in order to provide reserved airspace that allows space vehicles to travel through the American air traffic system.

But the organization is also studying the concept of a new international organization to regulate international space traffic. "The *International Space Flight Organization (ISFO)* [would be] an internationally-sanctioned organization that is the focal point for collaboration and information exchange for orbital or hypersonic point-to-point flights requiring international planning and notification to mitigate contention for airspace" [Reference 9]. This is an interesting concept that could be developed in more detail by governments currently supporting and/or financing RLV developments.

States will have to reach a consensus in the coming years on how they want to see RLV operations regulated internationally. It might be more cost-effective to integrate RLV issues in a revised ICAO "International Civil *Aerospace* Organization", where existing civil aviation structures could be adapted (i.e., specialized workgroups) and links with aviation issues could be easily created (i.e., air-traffic control procedures vs. orbital control).

Of course, creating a new international organization from scratch could bring a fresh start for space activities. However, not only expertise from the aviation world could possibly be missing in that scheme, but years of political and administrative hurdles — quite typical when it comes to the creation of a new international agency — could hinder efficient progress in regulating aerospace traffic issues.

## 4.    Conclusion

This paper has had the ambition to show that it is time to start debating how the future international aerospace regulatory environment should be drafted. Some might say that RLV technologies are still under development and it is too soon to elaborate plans. But, whether considering international air traffic perturbations or new sets of liability issues, future reusable launch vehicles will have to be integrated in — a yet to be written — international legal and regulation framework.

### References

1.  Reusability, the Key Word for the Future of Mankind in Space, *Spaceflight, Vol. 42,* February 2000
2.  Lafferranderie G. and Crowther D. (editors): *Outlook on Space Law Over the Next Thirty Years,* Kluwers Law International, The Hague, 1997
3.  Rey, R.: *Liability Issues and the Derivation of Reusable Launch Vehicle Space Worthiness Standards,* Independent Study Project, University of North Dakota, Grand Forks, December 4, 1999
4.  Collins P., Funatsu Y.: *Collaboration with Aviation - The Key to Commercialisation of Space Activities,* presented at session on Space Tourism of the 50th IAF Congress, Paper no IAA-99-IAA.1.3.03, Amsterdam, October 7, 1999
5.  FAA/DOT: License Application Procedures, *Advisory Circular 413-1,* August 16 1999
6.  COMSTAC, RLV Working Group: *Interim Report on RLV Licensing Issues,* p. 5, February 4, 1999
7.  Menter M.: Status of International Space Flight, *International Institute of Space Law,* 11, 1979
8.  ICAO Website: *ICAO Update July/August 1999,* www.icao.int/icao/en/jr/5406_up, April 20, 2000
9.  FAA-AST: *Concept of Operations for Commercial Space Transportation in the NAS (National Air Space) in 2005,* Version 1.1, January 14, 2000

# Report on Panel Discussion 4:

# How will the Legal Frameworks Need to Evolve?

J. Garget, P. Martinez, International Space University, Strasbourg Central Campus, Parc d'Innovation, Boulevard Gonthier d'Andernach, 67400 Illkirch-Graffenstaden, France

e-mail: garget@isu.isunet.edu, martinew@isu.isunet.edu

**Panel Chair : A. Kerrest, Faculté de Droit de Bretagne Occidentale, France**

**Panel Members:**

**M. Davis,** Attorney at Law, Ward and Partners, Australia
**R. Gress,** Manager, Licensing and Safety Division, FAA-OCST, USA
**C. Jolly,** Consultant, France
**F. Schroeder,** AAS Legal Counsel and Senior Associate, Milbank, Tweed, Hadley and McCloy LLP, USA

This discussion dealt with the legal frameworks, their limitations, and their foreseen evolution from both a public and a private sector viewpoint. During his introductory remarks, **A. Kerrest** highlighted the fact that the majority of countries do not have domestic laws regarding space activities, only an obligation to license their launch activities.

Several panelists stressed the need for new international regulations to fill the gaps in the current guidelines. These are needed in order to solve problems such as the assignment of liability when launching from the high seas, or the responsibility for space debris. As an example, **A Kerrest** mentioned the International Agency Debris Coordination Committee (IADCC), which intends to define rules that will deal with the problem of space debris.

Several speakers pointed out that the major space treaties are unlikely to change drastically, although some areas may need to evolve. **F. Schroeder** added that the treaties do not intend to regulate on a fine basis, but rather provide a broad international framework, and member states have a national responsibility to establish and maintain more specific domestic legal regimes.

The major contribution by the Australian government to the space activities of its country had been to establish a regulatory framework, and a department to deal with licensing and safety requirements. Within the

*M. Rycroft (ed.), The Space Transportation Market: Evolution or Revolution?, 227–228.*
© 2000 *Kluwer Academic Publishers. Printed in the Netherlands.*

government is a Space Industry Branch for science and resources, and a Space Launch and Safety Office. A governmental priority is to determine what assistance is required by the space industry in order to encourage the development of the commercial launch business, both nationally and internationally. **M. Davis** cited the active cooperation between Australia and Japan in the test landing of Japan's Hope spacecraft as an example of such international activity.

**R. Gress** stated that the time period of 180 days required by the Federal Aviation Administration (FAA) for the approval of licensing applications is not likely to decrease, due to the increase in launch activities and launch providers. However, it was not thought that it is currently necessary to create an international organisation for the coordination of RLV's.

Unlike the aviation industry, in which regulations have been developed through almost one hundred years of experience, there is no experience on which to base regulations for the currently undeveloped RLV industry. For this reason, only very general guidelines can be formulated at this time. These will ensure safety, yet maintain flexibility. Various environmental issues also have to be considered.

All the panelists agreed that space law must evolve slowly in response to technological developments, and that a revolution is not forseen.

# Session 5

# What is the Appropriate Role of Governments in Future Space Transportation?

Session Chair:

**F. Engström,** ESA, France

# An Industry Perspective of the Appropriate Role of the US Government in Future Space Transportation

J. Schnaars, Strategic Planning & Business Development, Reusable Space Systems, The Boeing Company, 5301 Bolsa Avenue, MC H014-C443, Huntington Beach, CA 92647, USA

e-mail: jayne.schnaars@boeing.com

### Abstract

It is in the public interest for governments to cultivate the technologies and provide legal/regulatory regimes that will encourage the development and safe operation of reusable launch vehicle (RLV) systems. The formulation and execution of government policies that encourage the economic success of the RLV industry will be crucial in enabling humankind to reap the benefits of a space-based market.

This paper explores the role of the US Government in opening the space-based market by addressing cogent payment structures, development partnerships, technology incubation, indemnification, and regulatory issues surrounding the development and operation of future reusable launch vehicles.

## 1. Introduction

The evolution of new transportation systems has opened new frontiers for human enterprise, enabling the creation of mass markets through which governments have reaped additional income through taxation. This income, in turn, has allowed new public initiatives to be realized.

The development of new transportation systems has, traditionally, required large amounts of government investment. For example, in 1926 the new Aeronautics Branch of the US Department of Commerce worked to expand navigational systems, monitored airfield operations, and evolved regulations to certify both aircraft and pilots for the nascent commercial aviation industry. These developments enhanced the willingness of insurance companies and lenders to participate in the rapidly developing aircraft industry, and were essential to the successful development of commercial aviation. Indeed, a carefully balanced framework of federal research programs and legislation represented a continuing pattern of influence and interaction that would benefit the commercial aviation industry in the years to come.

Today, a number of government/industry partnerships aimed at reducing space transportation development costs and risks are already in place or planned. The Evolved Expendable Launch Vehicle (EELV) program is just one example of such a partnership. The EELV program is focused on reducing launch costs by twenty-five percent (with a goal of 50%), while maintaining or improving reliability. With funding of $ 500 million for DoD-unique

M. Rycroft (ed.), The Space Transportation Market: Evolution or Revolution?, 231–242. ·
© 2000 Kluwer Academic Publishers. Printed in the Netherlands.

requirements and a separate launch services agreement from the US Air Force, Boeing is developing its commercially-derived Delta IV into a medium and heavy lift launcher family that will service both commercial and military markets.

In the area of space transportation, Boeing firmly believes that the size, requirements and credibility of the commercial market should determine which new systems are to be pursued by industry. We also recognize the important role that government will play in setting the stage for making the next generation of space transportation a reality. We must define the partnership between government and industry as they jointly pursue their respective public and private interests.

## 2.     The Case for Enabling a Commercial RLV Market

Tomorrow's space infrastructure will comprise a broad spectrum of commercial, military and civil space assets requiring launch services, on-orbit servicing and logistics. Today, many expendable launch vehicles service the commercial and military markets while human missions to Mir, the International Space Station and for low Earth orbit research are currently serviced by the Russian Soyuz and the US Space Shuttle system. Tomorrow's reusable launch systems must satisfy all of these market needs. But getting there will require a market driven solution that satisfies both commercial and government customers.

### 2.1   Market Enablers

Future commercial and government markets will demand reductions in the cost per kilogram to orbit, improvements in system reliability, faster turnaround times, shortened order cycles and high margins for crew safety.

After evaluating several Next Generation Launch System (NGLS) options, Boeing believes that the solution to these requirements and the successful commercialization of space lie in a Reusable Launch Vehicle (RLV). Fig. 1 illustrates the transition that must be made from ELVs to RLVs in the context of several key market requirements [Reference 1].

While we have a clear understanding of *what* the market enablers are, *when* the next generation RLVs will debut remains uncertain. With worldwide expendable launch vehicle capacity at its peak, expectations of enabling markets have dimmed in the wake of recent LEO communications systems business failures and the pervasive extension of the terrestrial communications infrastructure. Further, future RLV businesses face not only major technological and regulatory hurdles, but they must also seek a low-risk, investor-friendly business model that provides flexibility to address market uncertainties.

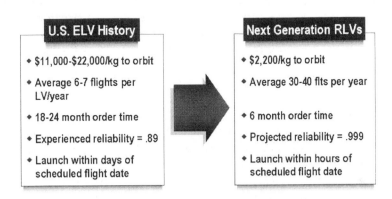

**Figure 1.** Key market requirements in the transition from ELVs to RLVs

Three key variables must be balanced in the decision to develop the next reusable space transportation system. The first variable is market need; the second is the investment requirement, and the third is the required rate of return on that investment.

Fig. 2 illustrates the *investment gap* between the market's investment requirements and industry's investment requirement. The only way to close this gap is to grow the market, reduce the investment or reduce industry's desired rate of return. Since industry is unlikely to accept below market returns, we will consider government's role in reducing risk and growing the market.

Commercial markets, rather than government, will clearly be the foundation of future privately owned and operated space transportation businesses, and *globalization* will be an integral factor in developing these new commercial markets. Still, as manifested by a 20 year forecast of today's traditional launch market segments [Reference 2], government revenues will continue to account for a significant amount of the total market, as noted in Fig. 3.

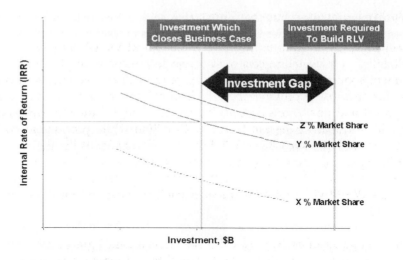

**Figure 2.** RLV investment gap illustrates market demand to close business case

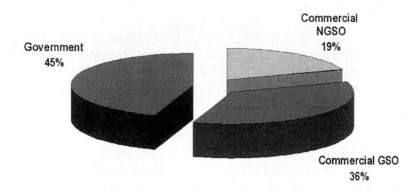

**Figure 3.** Addressable worldwide launch revenues

## 3.    Worldwide Space Insurance Thresholds

Launch insurance rate levels for commercial RLVs will largely depend on the reliability, replacement cost and projected income of the vehicle. The government sponsored flight technology programs discussed later in this paper will go far toward improving the RLV's reliability as well as reducing the development and operations cost of the vehicle. Still, future RLV development costs could fall in the range of $ 5 to $ 20 billion while production costs are likely to be as high as $ 1 to $ 2 billion per vehicle. With the theoretical worldwide space insurance capacity fluctuating between $ 1 and $ 1.2 billion and actual capacity of roughly $ 560 million, RLV asset insurance will be a major issue. Worldwide space insurance capacity allocations are shown in Fig. 4 [Reference 1].

The next generation RLV era will necessitate a paradigm shift for the space insurance industry.  Over this period, it will face dynamic changes in launch frequency, increased vehicle costs, second party liability resulting from commercial *human* missions and third party liability from routine launch and landings over populated areas. Furthermore, it must consider these changes in an international context.

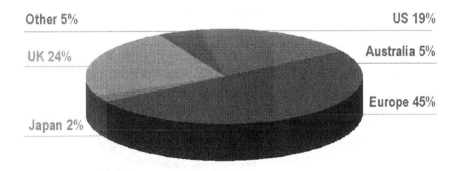

**Figure 4.** Today's worldwide insurance capacity

## 4.     The Role of  Government/Industry Partnerships

### 4.1   Reducing the Investment

Given a predicted market, industry will only commit to full-scale RLV development when there is a promise of an acceptable return. Therefore, reducing development costs and risks is essential; this also provides the best opportunity for industry and government to work together.

Other examples of government/industry partnerships involve NASA's Integrated Space Transportation Program (ISTP) [Reference 3]. Its goal is to develop technologies that enable new launch and in-space transportation systems with orders of magnitude improvement in safety, cost and reliability. This partnership includes the X-vehicle programs, whose specific objective is to reduce the payload price to orbit by an order of magnitude, from $ 22,000 to $ 2,200 per kilogram, within 10 years, and by an additional order of magnitude within 25 years. Advanced propulsion components, autonomous operations, new lightweight structures and innovative thermal protection system (TPS) solutions are just some of the technologies to be explored through these flight-test programs.

Management of these programs is handled through a contractual arrangement known as *Cooperative Agreements*. Cooperative Agreements permit industry and government to contribute complementary percentages of the negotiated development cost with the government's share of the cost being fixed. The contractor receives zero profit and recovers the government contribution only when agreed to milestones are achieved.

Several X-vehicle programs are underway, as depicted in Fig. 5. The Boeing un-piloted X-37 [Reference 4] will be the first NASA/industry reusable launch vehicle demonstrator designed to fly in both orbital and reentry environments, operating at speeds up to Mach 25. The X-37 is designed to demonstrate 41 airframe, propulsion and operations technologies aimed at significantly cutting the cost of space flight. The NASA/Boeing team will split the cost of the program, and the US Air Force has committed additional funding to demonstrate technologies needed to improve future military spacecraft.

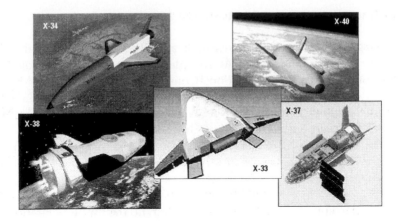

**Figure 5.** RLV flight technology demonstration programs will improve the
safety, cost and reliability of next generation RLVs

### 4.2  Loan Guarantees and Indemnification

It should be noted that there have been other methods of government
support proposed in the form of government subsidies. In 1998, Congressional
legislation was introduced with the goal of providing US Government-backed
loan guarantees to stimulate a more cost effective launch industry. If this
legislation is enacted, it will place NASA in the unique role of selecting which
RLV developers will receive these loan guarantees at below market interest
rates. It will also require NASA to cover, out of its very limited budget,
hundreds of millions of dollars in potential loan defaults. Boeing considers this
to be high-risk legislation that does not allow the development of the next
generation RLV to be driven by the commercial marketplace.

In a more positive light, legislation is pending in the US House and Senate
to extend third-party liability coverage to the commercial space launch industry
for four more years. Legislation is also proposed that will increase funding for
space-related industries and government to develop jointly a long-term plan
that addresses risk sharing. Boeing fully expects this legislation to be passed as
it is clearly prudent and necessary to ensure that the launch industry is
provided with a stable business environment in which to mature.

## 5.     Government-sponsored Incubation of RLV Technologies

In the US, decades ago the research activities of the National Advisory Committee for Aeronautics (NACA) played an important role in the development of commercial aviation. Considered to be a highly successful program, NACA was awarded six Collier trophies for its research. Of equal significance, however, is that most NACA research found application within the commercial aviation industry. Although NACA evolved into the National Aeronautics and Space Administration (NASA) in 1958, its legacy is invaluable for the technology transfer processes that can be applied to the reusable launch vehicle industry.

Today, the spirit of NACA is embodied in the implementation of the Integrated Space Transportation Program (ISTP) [Reference 3], the next generation of NASA/industry research partnerships. With its focus on technology development, NASA is committed to transitioning its routine space operations needs to the private sector. Instituted to define technology requirements and next generation launch system needs to meet future government and commercial needs, NASA is providing $4.5 billion of funding between 2001 and 2005 to resolve fundamental issues such as market timing, architecture and system solutions, commercial convergence, risk reduction and competition.

The ISTP will follow a three-part strategy, as noted in Fig. 6. The first part will be technical risk reduction, driven by industry needs that enable full-scale development of commercially competitive, privately owned and operated Earth-to-orbit RLVs by 2005 and operations by 2010. The second part will comprise the development of an integrated architecture with systems that meet NASA-unique requirements that cannot be economically served by commercial launch vehicles alone. The third part will address the procurement of near-term, path-finding launch services for select International Space Station requirements on commercial launch vehicles.

A follow-on to NASA's 1998/1999 industry-led Space Transportation Architecture Study (STAS) [Reference 3], ISTP is an excellent example of the role which the US Government should play in developing RLV-enabling next generation technologies, and RLV-sustaining technology and operability improvements. These areas of research will enable and continually improve the "robustness" technologies that will be fundamental to the certification of reusable launch vehicles.

Roles:

| NASA | Industry |
|------|----------|
| ◆ Basic research and generic high risk, high payoff technologies | ◆ Market forecasting and requirements definition |
| ◆ Mitigate risks for future systems | ◆ Commercial technologies & processes |
| ◆ Lead international efforts for human exploration | ◆ Commercial launch service solutions |

**Figure 6.** Commercial convergence for next generation RLVs requires NASA/industry partnerships

## 6. The Government's Role in RLV Operations within an International Law and Regulatory Framework

Interpreting how the current international air law and space law regimes will apply to the reusable launch vehicle conducting routine commercial operations will depend upon how quickly reusable launch vehicle technologies evolve. Second generation RLVs, because of their limited propulsion performance margins, will be required to use the most energy efficient, direct ascent trajectories to maximize payload mass fraction. And, because fuel reserves will be at such a premium, they will have extremely limited cross-range capability upon reentry, which will require optimal descent trajectories as well. Hence, the second generation RLVs will operate under the present legal regime in the same way as the Space Shuttle, with a few exceptions. For example, in the US, commercial second-generation RLV operators will be required to apply for launch licenses in accordance with FAA regulations, whereas the Space Shuttle, being a government sponsored program, will not.

As reusable launch vehicle performance capabilities and reliability improve with the advent of third-generation systems, it is conceivable that international commerce will begin to rely more and more on space transportation, resulting in routine commercial space operations *between sovereign nations* (e.g., space mission originates in one country, lands in another to refuel, takes on cargo, and returns to orbit, or to another point on Earth). It is at this point that the legal standards for an RLV will begin to diverge from those of the Space Shuttle, because the RLV will be capable of operating as freely within the atmosphere as it does in space. This enhanced technical capability will introduce a number of diverse "air law vs. space law" issues related to registration and marking, licensing and inspection, jurisdiction, traffic and transit rights, liability, and delimitation. And, because air law and space law are two different legal regimes (each with unique underlying principles), the regime in which the RLV is operating, relative to its mission profile, will have to be determined. This will be extremely critical in terms of settling any liability issues that may arise.

In air law, aircraft must meet airworthiness and vehicle certification requirements in accordance with national and international standards. These regulations are designed to promote safety and reliability in aircraft systems, and apply to all aircraft designs, whether they are operational or developmental. To enable operationally prevalent RLVs to achieve their maximum potential, they will also need to function within the confines of an international regulatory framework and an established airworthiness code. Boeing believes that second-generation RLVs will become catalysts for the codification of RLV-specific flightworthiness standards, and will validate these unique certification requirements through revenue-generating, operational flight-testing. The evolution of certified, commercial RLV operations within an international law and regulatory framework is depicted in Fig. 7.

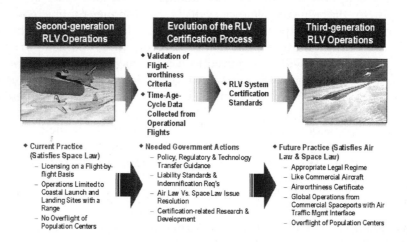

**Figure 7.** Evolving RLV operations within an international law and regulatory framework [Reference 5]

It is imperative to begin planning for the certification of third-generation reusable launch vehicles well before these systems are built and tested. First, such planning will initiate the process of removing a major source of uncertainty and risk from the marketing and financing end of developing these operationally pervasive vehicles. Secondly, it will enable the industry to work alongside government regulators and speed the process of certification. Finally, it will allow the legal and regulatory process to interact holistically with the physical design and manufacturing process of the vehicle. This, in turn, will influence the final design and operational characteristics of the system — directly affecting the indemnification rates and, ultimately, the business case for the RLV enterprise.

## 7. Summary

This paper has outlined a framework of government legislation and policies that will be conducive to developing the next generation of space transportation. Domestic government policies and legislation include launch indemnification, research and development support, technology transfer, flexible contract structures, and the implementation of regulatory standards. Fig. 8 outlines the process of how RLVs will evolve into market driven systems.

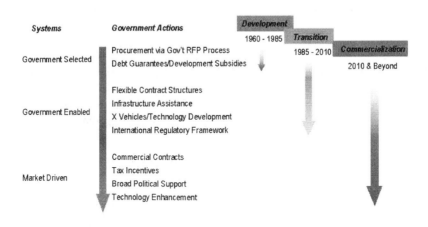

**Figure 8.** RLVs will evolve into market driven systems

A concept of international RLV flightworthiness regulation was also introduced to enable operationally pervasive third-generation systems to bridge the legal gap between air law and space law regimes. As flight-testing of second-generation RLVs continues, these systems will become catalysts for the eventual codification of RLV-specific flightworthiness standards.

The development of a next-generation reusable launch vehicle will be highly dependent on an ever-increasing global market for new and existing space services. This market, however, can only be enabled by a robust RLV industry which relies on a business incentive for government and industry to invest in the requisite technological advancements in propulsion, structures and avionics. By government and industry working together, the next generation of reusable launch vehicles will become a reality.

### References
1.     Stephens, R.: *How to Finance and Develop an RLV Industry*, presented at the First World Summit on the Space Transportation Business, Paris, France, May 11, 1999
2.     Federal Aviation Administration's Associate Administrator for Commercial Space Transportation (AST) and the Commercial Space Transportation Advisory Committee (COMSTAC): *2000 Commercial Space Transportation Forecasts*, May 2000
3.     Stephenson, A.: *US National Security, Civil and Commercial Space Launch: The Space Launch Initiative*, presented at the 3rd Annual Commercial Space Launch Forecast Conference, Washington, DC, February 8, 2000
4.     NASA Marshall Space Flight Center: *Pathfinder Program: X-34 and X-37 – Space Transportation*, <http://stp.msfc.nasa.gov/pathfinder/pathindex.html>. May 11, 2000
5.     Rey, R: *Deriving an Acceptable Level of Reusable Launch Vehicle Flightworthiness*, presented at the Space Technology Applications and International Forum (STAIF-2000), Albuquerque, New Mexico, February 1, 2000

# What is the Appropriate Role of Governments in Future Space Transportation? A European Industrialist Point of View

**P. Eymar,** AEROSPATIALE MATRA LANCEURS, 66, Route de Verneuil – B.P. 3002 – 78133 Les Mureaux Cedex, France

e-mail: patrick.eymar@LANCEURS.aeromatra.com

**Abstract**
Space launchers were developed initially under purely governmental funding for strategic and/or political reasons. Now the situation is evolving, the trend being to bring in more private funding in various ways.
AEROSPATIALE MATRA LANCEURS as a launcher designer and developer is especially interested in assessing the role of governments in the process. A recent European example and other in-house studies have shown that it is economically impossible for a private company to fund fully a new launcher; analysis of non-European approaches confirms this fact. European governments must therefore continue to play their supporting role along three lines: maintenance of the present competitiveness of existing expendable launchers, possible support of complementary expendable launchers if politically justified, and investment in new technologies to open the way to really cost-effective vehicles (reusable or not).

## 1.    An Initial Governmental Approach

Space launcher developments, in all countries but one, have been spin-offs of intercontinental ballistic missiles (ICBM). This initial move has had significant consequences that are still felt some half a century later. This is especially true for the relationship which exists between governments and launcher industries.

American, Soviet, British or French governments have borne totally the costs of developing these ICBM's and the first launchers for their own needs. Governments were at the same time the ones specifying the need, funding the development, buying the units and launching them with their own payloads. When a few years later commercial entities, at the time quasi-governmental ones, developed an interest into putting different kinds of payloads into space they were charged no more than the "run-on" cost or close to that. By no means were they asked to participate in the reimbursement of the development costs!

Once started, this process has been going on worldwide for decades, with different methods according to the countries concerned (explicit funding of development, "preferred" governmental rates, additional well-paid tasks, overhead re-funding, etc.). Nowadays some voices are raised to claim, or to request, that launchers (having reached their adult age) should exist without

*M. Rycroft (ed.), The Space Transportation Market: Evolution or Revolution?*, 243–248.
© 2000 *Kluwer Academic Publishers. Printed in the Netherlands.*

governmental support. In many cases these comments are coming from new space ventures projects, or start-ups and sometimes are heard by politicians always eager to cut unnecessary governmental expenses. The same voices may cry for government help a few months later, but the political damage has by then been done.

This appears to AEROSPATIALE MATRA LANCEURS, the main European launcher industrialist, as a dangerous move since it does not take into account the specificity of the launcher world.

## 2.   Specificity of the Launcher Industry

The launcher world is characterized by some very specific features:

- The limited serial effect: less than a hundred models are produced annually all over the world, with no more than 20 usually being from the same short series of production of the same vehicle

- The large ratio between non recurring and recurring costs (generally 20 to 40) for really new vehicles: if development costs have to be recovered, the launch prices would have to be doubled, or even trebled, killing the space business

- Extremely long development cycles (around a decade) requiring capital intensive development

- The peculiarity of the launcher world itself, which cannot be considered as a sub-part of a larger industry, but which requests a very large gathering of expertise with no other real applications

- The high technical standards requested, not balanced by strong profits

- The technology sensitivity which causes governments to scrutinise all international exchanges closely, and

- The political aspects biasing the market.

All these peculiarities do not today favour the interests of investors, as reflected in stock exchange figures. One of the major drawbacks is considered to be the long time that elapses before any return on investment may appear.

## 3.    Governments, the Largest Customers

It is of interest to identify who buys what presently, before trying to assess the future. AEROSPATIALE MATRA LANCEURS has been looking at all launches performed during the past three years (1997 to 1999) in order to extract the main trends. It is of course extremely easy to identify the launches; it requires some care to identify clearly which launches should be considered as commercial or non-commercial. It becomes more problematic to tackle launch revenues, since launch prices are not highly publicized. Fig. 1 shows the breakdown of launches per type of customer, the unit being the number of launches, Fig. 2, for the same breakdown, shows the income generated.

Around 35 launches have been performed over this period of time to place commercial payloads into orbit, while 45 to 55 represented government funded launches (with a ratio of roughly 60/40 between civilian and defense purposes). So 65% of launches are paid by governments. However, this figure is closer to 70% when considering prices (with very low prices estimated for CIS launches).

All this underlines the importance of the governmental sector for space launchers. These figures reflect the world situation, but not the European one which differs dramatically. Only 1 or 2 out of the 10 annual launches performed by Ariane vehicles are paid for by a European governmental entity, implying that this is not the way that European governments are supporting this industry.

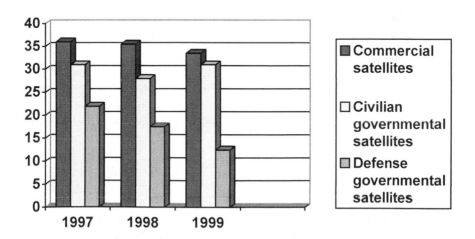

**Figure 1.** Number of launches per payload type (1997-1999)

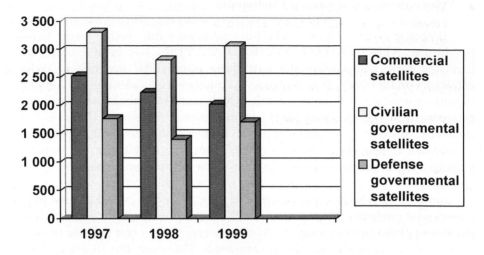

**Figure 2.** Generated income (in millions of $) from satellites launches per payload type

## 4.    Facing the Future

AEROSPATIALE MATRA LANCEURS's views on the future launch market (in the next two decades) rely more or less on the same sources used by other speakers at this Symposium. Our own understanding is as follows:

* The number of yearly launches will not increase globally

* The present rhythm of governmental launches will remain stable in numbers, but will represent a smaller share in terms of costs since the use of heavy vehicles will be reduced. If military satellites become less numerous, the development and use of space stations will compensate for the deficit

* The rate of launches for commercial entities will dwindle because substitutable technologies will take their share. In the last three years commercial launches were artificially high because of the satellite constellations launched (10 launches in 1997, 15 in 1998 and 12 in 1999); thus, the relative share of government funded payloads will increase

• The only way to recover a growing market should be to reduce the costs of access to space dramatically (by an order of magnitude?) enabling new uses to flourish.

So far no technical solution appears to us to be promising enough to be worth starting a full scale development, but work has to be done along this line.

## 5.    What Role for Governments?

When considering the coming years AEROSPATIALE MATRA LANCEURS expects a continuous and steady support from European Governments for their launching industry, but not exclusively for the long awaited goal which is the famous RLV. This support should also address other segments, such as the existing ELV's and the as yet non decided complementary (to Ariane 5) ELV's.

### 5.1   Support of Ariane 5

The present strong position of Ariane 4 in the segment of commercial satellite launches has been reached thanks to the past funding of all developments and to a continuous, though very limited, support covering enhancements. We expect the same situation to occur for Ariane 5 since there is no hope that more orders will come from the European states (as opposed to the USA or Russia) for large satellite launches.

### 5.2   Support of Complementary ELV's

Up to now there is no fully European launcher complementing Ariane 4 or 5. The existing Rockot or Starsem are answers to the market need provided by private industry moves without government funding (but with a strong and mandatory governmental political support at the time of implementation). Such an industrial involvement was possible because there were no large and lengthy investments to implement, clear business opportunities existed, and no reason of sovereignty had to be taken into consideration.

The situation is different if one considers that, in a given future not precisely known today, Europe will have a need for an indigenous launch capability to put aloft Europe's own military satellites (at a lower cost than using the heavy loader Ariane 5). No commercial reasons are sustaining this approach, so that full governmental support is then required. Depending on the political situation in the coming years, complementary launches will be of the small to medium class. The important point right now is to take all conservative measures which enable Europe to perform this later development at the minimum cost and, as far as possible, in the same move to improve the competitiveness of Ariane 5.

### 5.3  Support of RLV's

Two types of RLV's could/should obtain support from European governments:

- A manned RLV to replace the existing US Space Shuttle: since the main reason for a human presence in space is now linked to international endeavors such as the International Space Station, the common development of a follow-on to the Space Shuttle could be envisaged on an international basis, with purely public funding. The objective would be to reduce significantly costs (by 50%?) the present costs of the Space Shuttle

- A commercially oriented RLV (an order of magnitude cheaper than present ELV's): a very distant goal, private industry cannot invest in this. It is then the role of governments to pave the way for this new generation of space transport, which will open up new economic opportunities for society. Various mechanisms are possible and can be considered.

Not mentioned explicitly in the above paragraphs are all governmental actions, such as regulations, anchor tenancies, international rules of the game, etc.. These are, of course, either welcome or mandatory.

### 6.    Conclusions

Even if the launching industry has reached a mature status, it remains a fragile one and a non-classical one with respect to the usual economic standards; there is still a strong flavour of governmental involvement. This is caused by the size of the industry and more evidently by its political and strategic importance, non commensurate with the funds invested. It is also worth recalling the parallel which exists between access to space and the development of any publicly used infrastructure in terms of responsibility from a government point of view.

So the role of governments remains one of prime importance, not only helping industry to find new ways of space transportation, but also supporting existing industries which otherwise would rapidly decline, bringing with their fall large segments of other economic worlds.

# Evolving Government Roles in an Increasingly Commercial Space Transportation Market

**J. Rymarcsuk,** Business Development, International Launch Services, 12355 Sunrise Valley Drive, Suite 400, Reston VA 20191, USA

e-mail: jim.a.rymarcsuk@lmco.com

**E.E. Haase,** Strategic Planning, International Launch Services, 12355 Sunrise Valley Drive, Suite 400, Reston VA 20191, USA

e-mail: ethan.e.haase@lmco.com

**Abstract**
During the past few decades, commercial interests have established themselves in the space industry, becoming the largest segments in many space markets. The space transportation market has also taken part in this evolution of the space industry, with the introduction and growth of commercial launch service providers that serve increasing numbers of commercial customers. International Launch Services (ILS) is a leader in the transition to a more commercial industry. With the development of the space industry into a commercial marketplace, the role of government in space transportation has changed and will continue to change in the decade to come.

Governments are focused on three key roles within the space transportation market: use of space transportation services; development of technology and launch systems; and regulation of space transportation. The space transportation market has evolved from its inception as an industry fully sponsored by, used by, and controlled by government to an industry in which governments are special customers, development partners, and regulators. As the trend toward commercial services continues in the space transportation market, governments will evolve ideally into customers that are treated as any other commercial customer; partners that provide similar levels of development support worldwide; and regulators that promote commercial competition and fair trade practices.

## 1. Introduction: The Commercialization of the Space Industry

The space industry was born from government funding of technology development programs, leading to the launch of the first satellites and the first human spaceflight programs. Governments controlled the space industry and its direction for the first few decades of the industry's existence. This began to change in the 1970s and 1980s, as satellites became more common for telecommunications applications. As the market for satellite services began to grow, commercial companies began to develop plans that would ultimately lead to the growth of a commercial space industry.

The rapid growth in demand for telecommunication and information services worldwide in the past decade has been an engine of transformation for the space industry. Although the commercial space industry constitutes only

249

*M. Rycroft (ed.), The Space Transportation Market: Evolution or Revolution?*, 249–255.

about 5% of the telecommunications and information industry [Reference 1], demand for the specialized services that satellites deliver has propelled forward the need for commercial satellite capacity commercial revenue. As this industry segment grew, a commercial space transportation industry developed to provide access to space.

Government spending on satellites, launch vehicles, human spaceflight, scientific research, and other programs has been and still constitutes a significant part of the industry. While the commercial space industry continues to grow, however, government space spending has leveled off. Even in the United States, which has the largest space budget worldwide, spending has stayed relatively constant over the last several years.

The trend in increasing commercial space revenue and stabilizing government space spending is illustrated in Fig. 1 [Reference 1]. In 1997, commercial revenue in the industry outstripped government space spending for the first time. This trend is expected to continue as the demand for satellite telecommunications and information services will serve to increase commercial revenue over the next several years. During the same period, no appreciable increase in government spending is expected.

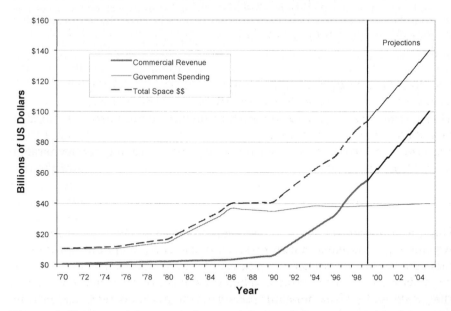

**Figure 1.** Commercial space industry revenue and government space spending, 1970-2005 [Reference 1]

## 2.    ILS: A Leader in Commercial Space Transportation

International Launch Services (ILS) is an example of an entity developed to serve the commercial space transportation industry. ILS is a joint venture of Lockheed Martin Corporation of the Untied States, and two Russian companies, Krunichev State Research and Production Space Center and RSC Energia. ILS was formed in 1995 by combining Lockheed Martin's Commercial Launch Services (which marketed the Atlas vehicle), with Lockheed Krunichev Energia International (LKEI), which was established in 1993 to market the Proton launch vehicle commercially.

ILS is a leader in the commercial space transportation market that has evolved along with the development of the commercial space industry. Although the United States government is one of ILS' key customers, launches for commercial customers have exceeded government launches for the last four years, as shown in Table 1.

|                      | 1996 | 1997 | 1998 | 1999 |
|----------------------|------|------|------|------|
| Commercial Launches  | 6    | 10   | 6    | 8    |
| Government Launches  | 2    | 2    | 3    | 2    |

**Table 1.** ILS launches, 1996-1999

## 3.    Evolution of the Role of Government in the Space Transportation Market

The heavily commercial space industry of today developed from an operation undertaken completely by government. As the commercial industry took shape, the role of government in the industry underwent a significant evolution. Government involvement in the space transportation market can be characterized in three key areas: space transportation use, technology/system development, and regulation.

### 3.1   User

The role of governments as users of space transportation has undergone perhaps the most noticeable transformation. In the early years of the industry, governments acted as the sole operator of space transportation services. Governments would procure launch vehicles from contractors, and conduct all launch operations for these missions. Government payloads were the priority for these launches, but other payloads were accommodated as the telecommunications industry began to use satellite capacity.

During the 1980s and 1990s, however, commercial launch service providers began to emerge to serve commercial satellite operators that were beginning to develop the satellite services market. Today, commercial providers dominate the market, serving both commercial and government customers. Commercial companies such as ILS and Arianespace provide the majority of heavy-lift launch services today.

Governments, once the sole providers of space transportation services, have generally taken on a reduced user role in the space transportation market. The U.S. government operates the Space Shuttle fleet through a commercial contractor and contracts for launch services from ILS and other providers for many of its mission requirements. ILS' Russian partner Krunichev provides similar services for the Russian government, while focusing on launching commercial payloads through ILS. While governments once controlled all space launch facilities, commercial spaceports have been developed in several regions, and the United States government now leases individual launch pads at its launch complexes to commercial providers.

## 3.2   Technology / System Development

Government involvement in the development of new launch vehicle technologies and launch systems has also undergone significant transformation. In the first few decades of the industry, governments directed and funded all aspects of vehicle and technology development. As commercial launch services' providers emerged, commercial concerns began to influence vehicle development. Government funding for new vehicle development began to decline, and commercial providers began to apply commercial funding to vehicle upgrades and new vehicle development. One recent example of such commercial development is the Atlas III series of vehicles. The Atlas III, which incorporates the Russian-developed RD-180 engine, was designed and produced using commercial funding primarily to meet the needs of commercial customers, who are demanding increased payload capacity from launch service providers.

Government launch system development support can have an impact on commercial competition as governments today provide uneven levels of support to launch vehicle programs. For example, development of the Ariane family of vehicles has been fully funded through the European Space Agency (ESA), while development of the next generation of launch vehicles for ILS and the Boeing Company (Atlas V and Delta IV, respectively) is funded jointly by the U.S. government and the commercial sector.

Countries with emerging space transportation capabilities (e.g. Brazil, India) are largely government supported. While this government support comes at a time when most of these nations are within the first or second decade of space launch capability, most of these countries have expressed a desire to enter the commercial space transportation market. As these newer launch vehicle programs develop and move into commercial applications, they will probably follow the same evolutionary path as described in this paper.

## 3.3  Regulation

The way in which the space transportation industry is regulated also changed as the market began to commercialize. As commercial launch services providers emerged, governments saw a requirement to develop regulations to govern the activities of launch service providers and their interactions with customers.

Vehicle safety, insurance and indemnification, and national security became key government concerns. In the United States, the Department of Transportation gained authority to license commercial space launches. This licensing authority focuses on ensuring the safety of the public during space launches, and providing for compensation for any personal or property damage associated with licensed launches. Commercial launches by U.S.-based launch service providers and the operation of launch sites by U.S. companies are subject to the Department of Transportation's licensing jurisdiction.

In 1999, following allegations of the transfer of sensitive technology to China, the U.S. government tightened its export licensing process (moving the authority for licensing the export of satellites from the Department of Commerce to the Department of State). All satellites built in the United States must receive an export license to be launched on a non-U.S. vehicle or launched for a non-U.S. customer. This process allows the U.S. to control tightly the export of technologies to other nations. Since 1993, the United States has entered into bilateral agreements with Russia, the Ukraine, and China that allow for the launch of western satellites on launch vehicles from these countries. These agreements serve the U.S interests of promoting commercial economic development in these nations while setting standards for average pricing that protect U.S. industry. Today, governments may also apply space transportation regulations to foreign policy goals. The United States government has applied the export licensing regulation and launch quota agreements described above to help meet its foreign policy objectives.

**4.     Future Evolution of the Role of Government in the Space Transportation Market**

As the commercial space industry continues its growth, the space transportation market will continue on its trend toward commercialization. This section explores government roles in the future that would best promote a near fully commercial space transportation industry, and the steps that should be taken by both government and commercial entities to reach these roles.

*4.1    User*

As the industry continues to become more commercial, governments should be served as commercial customers. Evolution to a system with little or no distinction between customer types will require changes from both governments and the commercial launch service providers.

In a fully commercial environment, governments would begin to act more like commercial customers both in the contracting process and in mission oversight. The government contracting process and the contract terms required by government customers must mirror the commercial process. While some government programs are already treated largely as international commercial customers (such as the Skynet system for the British Ministry of Defense), most government programs require contracting processes and contract terms more complex than those of commercial customers. Government requests for proposals may require a more detailed response than commercial proposals. To reach a more fully commercial market, commercial technical oversight also should replace the government supervision of launch operations.

At the same time, commercial launch service providers should adapt to serve the special needs of governments customers easily. Commercial providers must develop a level of schedule assurance and flexibility that satisfies government mission requirements. The broad range of performance requirements specified by government customers also must be met within the commercial launch vehicle offerings. In addition, security measures for sensitive payloads need to be in place.

*4.2    Technology / System Development*

The government role in launch technology and system development will evolve as the commercial space transportation industry continues to grow, but governments will always have some financial participation in the development of new launch technologies. Government support of technology research and development benefits both the launch industry and other government and national interests. As the commercial industry becomes more self-sustaining, however, government development funding will probably decline.

To promote fair commercial competition in the industry, the current disparities among governments in supporting their jurisdictions' space transportation companies should be resolved. Governments ideally would agree to provide fair and comparable levels of program support to the launch industry. This normalization of government support in the industry can be accomplished through international trade regulations and other multinational agreements.

*4.3   Regulation*

As commercialization of the space transportation industry continues, government regulation needs to focus on developing and promoting fair commercial competition. The goal of international trade rules should be to synchronize government regulation and provide a level playing field for all companies. Foreign policy concerns ideally would be removed from the regulatory process, as they disrupt the flow of commerce within the industry. Promoting a fair competitive environment will be difficult to accomplish with national interests embedded in the regulatory structure.

## 5.   Conclusion

The continued evolution to a more fully commercial space transportation industry will require changes in how both governments and commercial providers conduct business. Governments are moving from launch system operators to launch service customers, and should become more like commercial customers in the future. The development of new launch systems and new space transportation technologies was once exclusively directed and funded by governments, but in many areas this funding is declining as commercial launch providers begin to develop systems independently. The future role of government in funding vehicle development should be regularized across nations to encourage a fair competitive environment. Governments have established a regulatory structure for the commercial space transportation industry as it has developed, but the future regulatory regime needs to focus on ensuring fair international competition.

**References**
1.   Futron Corporation: *Satellite Industry Guide,* Futron Corporation, Bethesda, MD, 1999

# Government Intervention in the Commercialization of Launch Services: Japan and Europe

K. Suzuki, College of Policy Science, Ritsumeikan University, 56-1 Tojiin-Kita machi, Kyoto, 603-8577, Japan, and Sussex European Institute, University of Sussex, Falmer, Brighton, BN1 9SH, UK

e-mail: kazutos@yifan.net, ksv20084@sps.ritsumei.ac.jp

### Abstract

This paper aims to provide an explanation of the Japanese attempt to commercialize the launch service of H-IIA and of how the government played a role in the process, compared with European experiences. Although Japan and Europe share a similar objective — not to depend on American launcher technology — their policy rationales for developing launchers are quite different. The Japanese objective was driven by the resentment of engineers who had to accept American leadership in launcher developments since the end of the 1960s, whereas the European objective was driven by the traumatic experience of Symphonie. The Japanese launcher development policy in which engineers have a strong influence was therefore driven toward an entirely domestically produced launcher concept. Whereas European governments created Arianespace even before the first successful launch of Ariane 1 to compete with American launch services, the Japanese government's priorities were to develop technological capabilities of its domestic industry with less intention to commercialize the launch service. The Japanese decision to create the Rocket System Corporation (RSC) and to commercialize H-IIA was, in fact, taken because of the accomplishment of the primary goal of 'domestically produced launchers.' However, the process of the commercialization of Japanese launchers has faced a serious problem in recent years. Successive failures have not only cast doubt on the potential and reliability of H-IIA and forced the RSC to set back its business, but also criticisms towards J-1 programs and NASDA management have fuelled the debate on the objectives of Japanese launcher development. Apparently, Japan seems to be willing to continue its commercialization process even after the failures, but it may be quite difficult for RSC to be as successful as Arianespace.

## 1. Introduction

Although most existing launch services are operated under private entities, those launchers were initially developed by governmental funding with particular political purposes. Obviously, the primary purpose for developing launch vehicles is to get access to space, but each government had different objectives to have access to space. Likewise, each government has its own rationale to commercialize launchers. The process of commercialization, therefore, should be regarded as part of a government's strategy to transfer its competence to private companies.

This paper seeks to analyze the intentions and objectives of governments to commercialize their launch services through a comparison of European and Japanese launchers. The process of commercialization as a government strategy,

257

*M. Rycroft (ed.), The Space Transportation Market: Evolution or Revolution?*, 257–265.
© 2000 *Kluwer Academic Publishers. Printed in the Netherlands.*

despite its similar appearance, should be regarded as a different process according to its history and objectives.

## 2.    Contrasting Historical Developments

### 2.1   *Japanese Launcher Development*

There are five distinctive features at the initial stage of Japanese launcher development. First, Japan, like Germany, was banned from any research related to aeronautical technology under Allied occupation until 1952, but a handful of scientists energetically continued their studies of rocket technology. Scientists, instead of government, were the leading force to promote rocket development. The Japanese government played an indirect and rather minor role at this stage, such as to support academic activities through the Ministry of Education (MoE). Following that, the second distinctive feature is that the launcher development solely took place at Tokyo University (TU). The role of TU in the post-war space development has been widely discussed [References 1,2], and launchers were not an exception. The monopoly of launcher technology by TU (apart from specific technological competence in the National Aerospace Laboratory) left the Japanese government unaware of the potential of launching applications satellites. However, influenced by the success of the Apollo program and Intelsat, the Japanese government finally began to have an ambition to develop applications satellites and its own launcher. The newly established Science and Technology Agency (STA) was entrusted to develop application technology whereas TU and MoE continued their scientific programs. Obviously, such a division of resources and manpower was not efficient to catch-up advanced launcher technologies, but TU and MoE stubbornly rejected the idea of integrating their launcher sections under STA's control. This late governmental intervention into space applications and bureaucratic divisions and conflicts continue today as the third distinctive feature of Japanese launcher development. Fourthly, the Japanese application launcher developed under STA was significantly influenced by American technology. The Johnson administration, as a part of package agreeing to return the sovereignty of Okinawa, offered to provide technical assistance to Japan to maintain the American leadership, which was seriously damaged during the Vietnam War and by trade frictions [Reference 3]. The Japanese space agency (which later became NASDA) was able not only to save R&D costs, but also to obtain access to state-of-the-art technology. Such a favorable offer for Japanese space development notwithstanding, many scientists and engineers were not happy with the agreement, since it would potentially destroy their past work within TU. For them, accepting American technology and engineers to develop their national launcher appeared to be a humiliation and a serious blow for the endogenous technological development. Thus, finally, the concept of achieving 'a 100% domestically produced launcher' was added as a strategic objective for Japanese launcher development, and eventually it became the priority for the entire space development program. Thus, NASDA launchers (N and H series) were developed with great emphasis on increasing the percentage of domestically produced components, and the current series (H-II) was clearly targeted to achieve 100% domestic production. Although NASDA achieved its primary objective, there were concerns over cost-efficiency and the potential for future commercialization was largely neglected by policy-makers.

## 2.2   *European Case*

The history of European launcher development was also a very complicated and confusing one. Unlike the Japanese case, European governments, particularly in Britain and France, began launcher development at the national level, based on military programs. However, the concept of a 'European' launcher emerged at a very early stage, largely because of historical contingency that the British government changed its defense strategy to dispose of its IRBM (Blue Streak); it decided to Europeanize its technology to save money and to share the cost of transforming missile to launcher. But, the institutional framework for launcher collaboration, European Launcher Development Organization (ELDO) was critically weak to manage a complex production system because the Member States claimed to protect their own technologies and competencies. In other words, European governments neither had a clear and coherent objective nor the institutional framework to develop a reliable launcher. However, the situation dramatically changed when the French government recognized that the US government would not allow the launching of the Franco-German applications satellite (Symphonie), which was potentially threatening the American monopoly of emerging profitable commercial space activities. The French government, which has proposed a new launcher under its leadership, realized that the lack of autonomous access to space was a life-or-death problem for European space industries and agencies, and encouraged its partners to support new launcher program [Reference 4]. This was by far the most important change in the history of European space development. Since then, the French government and its space agency (CNES) took the leadership to make a coherent institutional structure for launcher development, which overcame the problems associated with the intergovernmental nature of European collaboration, and to set the objective to establish autonomous access to space.

Unlike the Japanese case, European policy-makers shared a sense of the urgent need to attract customers for Ariane from the beginning, largely because of its competition against American launchers. European governments, particularly French one, were keenly aware of the shortfall of demand within European and national governments to justify Ariane development, so that their primary target was set to join the competition to launch international applications satellites such as the Intelsat series [Reference 5]. However, European governments were not in consensus on the procedure of commercialization. The French and German governments demanded to move on to commercial production to recover the development costs which they bore. Meanwhile, other Member States regarded ESA as a research and development agency, and they were also concerned at the risk and cost of producing a launcher for commercial purposes without any concrete orders from customers. Being impatient and frustrated, the Director of Launch Vehicle of CNES, Frédéric d'Allest, proposed in December 1977 a solution which was to separate commercial activities from the functions of ESA, and to set up a commercial company under the close supervision of space agencies and industry [Reference 6], which later became Arianespace. Against the American Atlas-Centaur and Space Shuttle, Arianespace made a bid to launch Intelsat V in the early 1980s, and was selected in 1978, even before the first experimental launch.

### 2.3   Historical Comparison

There seem to be three major factors which made a difference to the policy rationales for launcher development and commercialization in Japan and Europe. First, the relationship with the United States played a crucial role to divide the paths of launcher development. The Japanese space authority did not have difficulties to access American launchers, largely because the Japanese space development began relatively late, and therefore NASA and the American space industry did not have to worry about competition. Furthermore, the American authorities regarded Japan as a junior partner of space development to which transferring obsolete technology could be used as a diplomatic card. On the other hand, competition in the arena of telecommunications satellites between Europe and the US in the late 1960s had circumscribed the policy options for Europe to develop a launcher of its own. Secondly, there was an interesting paradox of launcher availability. On the one hand, Japan had already successfully launched its own launcher but used purely for scientific purposes and, on the other, Europe had failed in all its attempts to launch satellites into orbit, both for scientific and application purposes. However, Japan chose to depend on a American launcher whereas Europe could not. This can be interpreted that Europe was in a more vulnerable position vis-à-vis the United States than Japan in terms of a practical solution for the crisis of Symphonie. Europe could not turn to Russia (then the Soviet

Union), the only other country that had launch capability, in the height of the Cold War.

Finally, there was a significant difference for the objective to develop launchers. For Japan, although the major purpose was to launch applications satellites, it was the achievement of high technological standards and the catching up with advanced countries that became important objectives for the launcher. Therefore, there was no immediate need to justify its development and the production of launchers by attracting customers outside Japan (there was also an agreement with the US to limit the use of launchers in the early phase). On the other hand, European governments were keenly aware of the necessity to attract customers not only to establish a competitive position against US launchers but also to recover the costs and to justify the public spending. This difference is deeply rooted to a fundamental question: what is space 'industrial policy'? The Japanese space industry is not purely an 'aerospace industry'. Japanese space companies operate their space business alongside with other industrial sectors and, therefore, space has been regarded as an additional activity to add fame and pride to the companies. But, in Europe, space industries are a significant proportion of their 'aerospace and defense industries'; the governments have strong incentives to support the industry but they were not able to provide sufficient size of public procurement at both national and European levels to sustain the competitiveness of the industry. Thus, the commercialization became the central issue for the launcher development.

## 3.    Policy for Commercialization in 1990s

### 3.1    Japanese Case

The turning point for Japanese development came when its primary objective — a 100% domestically produced launcher — was accomplished. The success of H-II gave a strong confidence to Japanese engineers that they had achieved the goal to join the club of 'advanced space countries'. However, they were also thrown into the situation where technological achievement was no longer a justifiable political objective. The idea of commercialization, which gradually became the major game in town, seemed to be a suitable objective to convince Japanese politicians and, moreover, the Ministry of Finance. In fact, the idea of developing a new launcher for commercial purposes emerged not from the political or senior administrative levels, but from the young engineers who were engaged in the H-II development. While they were developing H-II, they felt that there were unnecessary costs, associated with the concept of '100% domestic production', and that it would be possible to reduce the production costs to the level of other commercially operating launchers. The findings of these young engineers were soon to be taken up by NASDA managers and STA to support the rationale for developing the next generation launcher, H-IIA [Reference 7].

However, the concept of commercialization was not a stranger to the Japanese space authorities. When NASDA began to plan the H-II program in the early half of 1980s, there was serious effort to make the new launcher competitive in the global launch market. Some people in NASDA were aware that domestic public and private users began to show interests in using foreign launchers. Although the idea to make H-II competitive was a difficult mission to complete, largely because of sharp rise of exchange rate, which made H-II price almost doubled, and also because of the priority to develop domestic technological capability, NASDA established a commercial company, Rocket System Corporation (RSC), which carries out the responsibility for production and marketing of H-II. The RSC, imitating Arianespace, was funded by 73 aerospace (heavy industry and electronic) companies and banks without the financial involvement of NASDA or of government. The outline of RSC gives an impression that it is purely a commercial company that would be aggressively competing in the launch market. However, the reality looked rather different. Apart from Ministry of International Trade and Industry (MITI) which is the most enthusiastic supporter for launcher commercialization but less powerful in decision-making for space development, the government and NASDA did not prioritize launcher commercialization, when the RSC was established, instead of the '100% domestically produced launcher' concept. This unfortunate timing of the establishment of RSC − well before the development of H-IIA − gave completely different characteristics to RSC as a replacement of the role of NASDA for launcher operation, because its product (H-II) was far from being competitive, and as a result, the RSC was not able to develop a commercial strategy. (In fact, all private satellites have been launched by foreign launchers since the establishment of RSC.) Furthermore, the consecutive failures of H-II forced the schedule of launcher development to be changed, and cast doubt of the prospect of reliability of H-IIA, which significantly constrained the strategic option for RSC [Reference.8].

Furthermore, what I call 'J-I saga' in 1998 added confusion to the process of commercialization. J-I, a small launcher program, developed under a 'partnership' between NASDA and ISAS, was to launch a 1-tonne commercial satellite into LEO using a H-II first stage and M-IIIS second stage. The program was targeted to enter the commercial market for small satellites by using off-the-shelf technology, but the development costs seemed to be more than double those of foreign launchers. Going from bad to worse, the demand for small launchers was not strong in both commercial and public market, so that there has been only one launch since 1995. And, worst of all, the Space Activities Committee (SAC) published a report in 1998 that the operational cost of J-I was too expensive and the estimated demand was not actually reflecting reality. It also criticized the 8-year gap between the first and second launch which seemed to be a big waste of budget and resources [Reference 9]. The report was immediately followed by a more critical report from the Management and Coordination Agency (MCA), a non-ministerial agency under the Prime Minister's office, to oversee the procedure and efficiency of government ministries. It bluntly criticized the bad cost-efficiency performance of J-I and the overestimation of demand and cost, and it recommended halting the program if it was not possible to reduce operational cost to the level of a similar foreign launcher [Reference 10]. The intervention by SAC and MCA represents the emergence of cost-consciousness among government agencies, on the one hand, but it also reflects the lack of competence of NASDA and ISAS for commercial-oriented programmes.

## 3.2   European Case

The success of Ariane raised great concern for US launcher developers and service providers. Losing out to Ariane for the Intelsat V contract was not only a significant loss of a prosperous customer, but also a loss of the monopoly of the launcher market. The American industry and government began to mount a legal challenge to Ariane's operations. Among other allegations from the American side, the case raised by Transpace Carriers Inc (TCI), an American private launch service company established in 1984, was an epoch-making case. In the case file, TCI insisted that Arianespace was subsidized by European government in several ways: (1) a two-tiered pricing system whereby European customers were charged 25-33% more per launch than export clients, (2) use of the Kourou facility for an unreasonably low price, (3) government subsidized mission insurance rates which would otherwise be passed on to the client [Reference 11]. However, it was obvious that the American government could not take any immediate action against Arianespace because of its practice of subsidizing the Space Shuttle. The TCI petition was filed to the US Trade Representative (USTR) under Section 301 of the Trade Act. ESA and USTR, representing the European and American governments respectively, held several intergovernmental consultations to settle the matter, but the meetings were not successful because of the differences of understanding of the meaning of consultation. For ESA, the meeting was considered to be independent of the Section 301 procedure and the opportunity to negotiate the practice of ELV and Shuttle sales, whereas USTR used the meeting as a fact-finding process within the procedure, which required ESA to open up its pricing policy. While the meeting did not produce any agreement or guideline for launch services, the US President acknowledged that the ESA practice was not substantially different from that of the US, and, therefore, that it would jeopardize the position of the US government if it took action against Arianespace [Reference 12]. Nevertheless, it was an important decision for European governments, ESA and Arianespace to establish a level playing field with American launchers.

Thanks to the US policy to concentrate on the Space Shuttle as the launch service vehicle, Arianespace marched on the commercial market to increase its market share. However, the changes in the international environment in the 1990s, particularly the collapse of the Soviet Union, posed a concern for Arianespace. The European governments, as well as the US government, were afraid of competing with former Soviet launchers, which are extremely cheap and relatively reliable, coming into the international (more precisely Western) launching market. In the early 1990s, a team from the European Commission (represented by DG I), ESA and Arianespace, and the Russian authorities achieved an agreement to set a quota for launching commercial satellites by former Soviet launchers. However, when the European Commission brought the agreement to be authorized by the Council of Ministers, France vetoed the agreement on the basis that France is responsible for an intergovernmental agreement on matters of a French registered company in space sector, and that the Commission has no competence in this matter. France, although it took a position in favor for a quota at the ESA level, rejected the agreement at the EU meeting, simply based on national prestige and in order to maintain control over space activities on national level.

## 4.     Conclusion

Although the process of the commercialization of Japanese launchers seemingly follows in the steps of Arianespace, there are a number of fundamental differences in the objectives of European and Japanese policies and their history. First, there is a fundamental difference in policy objectives. Although Europe and Japan share a similar objective — to gain autonomous access to space —, the rationale behind them is quite different. This, in fact, made a big difference in the 1970s whether to commercialize new launchers. Secondly, the institutional characteristics make a significant impact on the ways in which Japan and Europe decide their programs. In the case of Japan, the decision-making authority is widely dispersed among ministries, which have different policy norms for commercialization, although the programs are coordinated at the level of the SAC. Particularly, the dominance of STA and weakness of MITI in decision making for space policy shape the attitude of the Japanese government for commercialization. On the other hand, European governments, despite their different interests, are generally in accord to promote and support the commercialization of its launchers. Since commercialization became a more important issue for space agencies, they are now engaged in a process to strengthen the relationship with the EU, particularly for policy coordination and international negotiation.

Apparently, Japan seems to be willing to continue its commercialization process by concentrating on H-IIA developments even after the H-II failure. However, the outcome of commercialization would be different from the European case as long as the Japanese authorities, particularly NASDA, STA and SAC, keep emphasizing technological issues and lacking commercial management capabilities.

### References

1.  Saito, S.: *The story of Japanese Space Development: The Dream of Harbingers for the Domestic Satellite* (Nihon Ucyu Kaihatsu Monogatari: Kokusan Eisei ni Kaketa Senkusya tachino Yume), Mita publisher, Tokyo, 1992
2.  Godai, T.: *H-II Launch Vehicle Lift Off* (Kokusan Roketto H-II Ucyu heno Cyousen: Saisentan Gijyutsu ni Kaketa Otoko tachino Yume), Tokuma Shoten, Tokyo, 1994
3.  McCurdy, H.E.: *The Space Station Decision: Incremental Politics and Technological Choice*. The Johns Hopkins University Press, Baltimore, 1990
4.  Carlier, C. and Marcel G.: *The First Thirty Years at CNES: The French Space Agency 1962-1992*. La Documentation Francaise, Paris, 1994
5.  Chadeau, E. (editor): *L'Ambition Technologique: Naissance D'Ariane*. Editions Rive Droite, Paris, 1995
6.  d'Allest, F.: Why is the Most Widely Used Launcher in the World European?, *Proceedings of Twenty Years of the ESA Convention in Munich, 4-6 September 1995: ESA-SP-387*, pp. 47-50, 1995
7.  NASDA: *I am the Controller of H-IIA Development*, <http://spaceboy.nasda.go.jp/spacef/sp/j/02_imoto.html>.     NASDA Spaceperson, March 21, 1997
8.  Nishiyama, K.: *Hughes Deals Crippling Blow To Japan's Rocket Program*, <http://www.spacedaily.com/news/japan-hughes-00a.html>. Spacedaily, May 25, 2000
9.  Space Activities Committee, *Report on the Evaluation of Space Transportation*, June 1998 (Japanese only)
10. Management and Coordination Agency, *Report on the Administrative Investigation on Space Development*, April 27, 1998 (Japanese only)
11. Krige, J.: The Commercial Challenge to Arianespace: The TCI Affair, *Space Policy, Vol.15 no.2*, pp.87-94, 1999
12. van Fenema, H. P.: *The International Trade in Launch Services: The Effect of U.S. Laws, Policies and Practices on Its Development*. University of Leiden, Leiden, The Netherlands, 1999

# International Trade in Commercial Launch Services: Adopting the World Trade Organization General Agreement on Trade in Services (WTO/GATS)

T.C. Brisibe, Le Goueff Avocats, 9, Avenue Guillaume, L-1651, Luxembourg

e-mail: LE_GOUEFF@vocats.com

### Abstract

Commercial launch services, once the exclusive domain of rocket manufacturers using government launch facilities, have evolved into an international business. This paper addresses the conflict between national security concerns and the stimulation of trade in commercial launch services which constitute the result of utilizing a dual-use technology, resting on the same industrial base as military applications and subject to non-binding international accords on export control.

Despite fora through which governments discuss export control issues such as the Missile Technology Control Regime and the Wassenaar Arrangement, there remains great diversity in the constitutional mechanisms, legal systems, systems for administration, industrial framework, foreign policies and economic goals in different countries. The object of this paper is to examine current international guidelines, vis-à-vis selected national export control policies, influencing the cross-border provision of launch services, in the context of the WTO/GATS, with a view to recommending a regime that will foster true internationalization of the commercial launch market.

## 1.    Introduction

Space transportation systems ("launchers") have been described as being used to carry a payload from point A to point B. This transport, which may be from a planet to orbit, orbit to planet, orbit to orbit, planet to planet or between places located on the same planet [Reference 1], possesses a military-strategic and national security origin. Although since the mid-1990's the commercial use of launchers ("launch services") has grown in response to an increase in global demand, it has been contended [Reference 2] that there is a general agreement amongst experts and policy makers alike that it is impossible to distinguish between peaceful space launch technology and offensive missile technology. In other words, only one aspect of a rocket — the payload — differs between civilian satellite launchers and nuclear weapons-tipped missiles [Reference 3].

Considering *inter alia* the need to guard against the transfer of sensitive dual-use goods and technologies, the Wassenaar Arrangement on export controls for conventional arms and dual-use goods and technologies (the "WA") was established after receiving final approval by 33 co-founding countries in July 1996 [Reference 4]. For similar reasons, in April 1987, a number of countries had confirmed compliance to a common international export policy known as the Agreement on Guidelines for the Transfer of Equipment and Technology related to Missiles (the "MTCR") [Reference 5]. Both international accords are

267

*M. Rycroft (ed.), The Space Transportation Market: Evolution or Revolution?, 267–273.*

non-binding and their implementation remains the responsibility of the signatories, through their national laws and policies.

In 1994, 125 nations signed the WTO Agreement, which achieved *inter alia* the completion of The General Agreement on Trade in Services (GATS) [Reference 6]. The GATS is the world's first multilateral agreement on investment, since it covers not just cross-border trade but every possible means of supplying a service, including the right to set up a commercial presence in the export market [Reference 7]. In today's launch services industry [Reference 8] there are 19 launch systems from 6 countries, using 14 geographic sites [Reference 9]. Nonetheless, unlike other forms of competitive economic activity where there is an availability of the best product at the best price, to be determined by free market forces, it has been contended that in launch services there is no competition in spite of the availability of a variety of launch vehicles [References 10, 11].

Section 2 of this paper discusses the objective/scope of the WA and the MTCR; Section 3 highlights the French and Ukrainian export control policies; Section 4 outlines the principles of the GATS and suggests the possible application of the GATS to the liberalization of launch services, similar to measures taken in the global telecommunications industry that in 1997 generated revenues, which stood at US $ 644 billion, with investment totaling US $ 170 billion [Reference 12]. Section 5 concludes the paper with recommendations.

## 2.     The Wassenaar Arrangements and MTCR

The purpose of the WA [Reference 13], as reflected in its Initial Elements, was to contribute to regional and international security by (i) promoting transparency and greater responsibility with regard to transfers including *inter alia* dual-use goods and technologies, (ii) ensuring that transfer or diversion of these items do not contribute to the development or enhancement of military capabilities which undermine these goals, (iii) complementing the existing control regimes for weapons of mass destruction and their delivery systems, as well as other internationally recognized measures designed to promote transparency and greater responsibility, and, (iv) enhancing cooperation to prevent the acquisition of armaments and sensitive dual-use items for military end-uses. The arrangements are not directed against any state or group of states, nor intended to impede *bona fide* civil transactions. Participating States were enjoined to control all items set forth in the List of Dual-Use Goods and Technologies [Reference 14] and the Munitions List [Reference 15] with the ultimate decision to transfer or to deny a transfer of any item being the responsibility of each participating State.

The Missile Technology Control Regime (MTCR) is an informal and voluntary association of countries sharing the goals of non-proliferation of unmanned delivery systems for weapons of mass destruction [Reference 16]. It seeks to coordinate national export licensing efforts through common export policy guidelines (the MTCR Guidelines) applied to a common list of controlled items (the MTCR Equipment and Technology Annex). All MTCR decisions are taken by consensus. Individual partners are responsible for implementing the Guidelines and Annex [Reference 17] with sovereign national discretion and in accordance with national legislation and practice. MTCR controls are not intended to impede peaceful aerospace programs or international cooperation in such programs nor are they designed to restrict access to technologies necessary for peaceful economic development.

## 3.    National Export Control Policies

### 3.1    France [Reference 18]

According to the policy statement on export controls for *inter alia* dual-use items submitted to the WA [Reference 19], French export controls were introduced by a decree of 1944. The export controls have been implemented in accordance with rules defined by a European regime established by the EU Regulation 3381/94 of December 1994 and the Common Foreign and Security Policy decision 94/90 of December 1994 (as amended). The policy confirms France's acceptance to comply with additional controls, and guarantees that the control of exports of goods that are not military equipment but whose export it regards as extremely important is conducted by means of a national licensing system in accordance with the European criteria [Reference 20]. The policy concludes by stating that..."all actions conducted by France in the field of exports of arms and dual-use goods and technologies"... "take place within a regulatory and political framework incorporating the laws and obligations agreed at European and International level".

### 3.2    Ukraine [References 21, 22]

The Ukraine, which is party to the WA and MTCR, does not offer a uniform law to provide legal authority for national export controls. The legal basis for the export control system is contained in decrees by the president, which formulate elaborate procedures for State Export Controls [Reference 23], and re-organized the procedures and institutions, which establish both the arms export and control policies [Reference 24]. The controls are applied to a national list [Reference 25] divided into five categories, which includes *inter alia* dual-use goods. The controls include licensing procedures supervised by the State Service on Export Control that works in tandem with consultation procedures of the Commission on Export Control Policy and Military and Technical Cooperation with Foreign Countries [Reference 26].

## 4.     The World Trade Organization General Agreement on Trade in Services [Reference 27]

The General Agreement on Trade in Services (GATS) is the first ever set of multilateral, legally enforceable rules covering international trade in services. It consists of 29 (twenty nine) articles, 8 (eight) annexes and 130 (one hundred and thirty) schedules of commitments on specific service sectors. The agreement operates on three levels [Reference 28]. In particular, the schedules of services are the means by which signatories make legally binding commitments on Market access (Article XVI), National treatment (Article XVII), and may contain Additional commitments (Article XVIII) that create an open-ended possibility to negotiate commitments on measures affecting trade in services that are not captured by market access and national treatment. Additional commitments for the telecommunications sector were embodied in a document known as the "Reference Paper" [Reference 29].

Whether full or limited access is granted, members may not take measures that reduce the level of access inscribed in their schedules. It has been contended that..."the ensuing liberalization of global telecommunications through the opening of national markets to international competition has gone hand in hand with a convergence of domestic telecommunications companies with those of other nations to form multinational alliances, in order to enlist additional capabilities, create synergies and share the risks and huge costs involved. The result of this regulatory and strategic revolution is a phenomenal growth of the telecommunications industry" [Reference 30].

## 5.     Conclusions and Recommendations

The application of the aforementioned controls tends to raise international political problems as non-participants may see them as unjust. Their implementation through non-uniform/heterogeneous national laws and policies may result in the creation of non-competitive market conditions, which restrict rather than develop the launch industry. Dr. Johnson-Freese agrees that a start be made towards developing an international, comprehensive, rational approach to technology transfer and export controls. New, clear rules appropriate for real world conditions are needed, and a regulatory process appropriate to enforce those rules [Reference 31]. It is in this context that the GATS procedures, which underscore the rule of law, are deemed appropriate.

### References

1.  Houston, A., and Rycroft, M.: Keys to Space: *An Interdisciplinary Approach to Space studies*, McGraw-Hill, 1999
2.  Fenema, H, P van.: *The International Trade in Launch Services: The Effect of U.S. Laws, Policies and Practices on its Development*, The Netherlands, 1999
3.  The Salt Lake Tribune: *CIA Says Chinese Satellite Launchers, Missiles* Similar, <http://www.sltrib.com/1998/may/05221998/nation_w/3472.htm>. May 22, 1998
4.  The participating States of the Wassennar Arrangement are: Argentina, Australia, Austria, Belgium, Bulgaria, Canada, the Czech Republic, Denmark, Finland, France, Germany, Greece, Hungary, Ireland, Italy, Japan, Luxembourg, Netherlands, New Zealand, Norway, Poland, Portugal, the Republic of Korea, Romania, Russia, Slovakia, Spain, Sweden, Switzerland, Turkey, Ukraine, United Kingdom, and the United States
5.  Current MTCR membership includes: Australia, Austria, Argentina, Belgium, Brazil, Canada, Czech Republic, Denmark, Finland, France, Germany, Greece, Hungary, Iceland, Ireland, Italy, Japan, Luxembourg, Netherlands, New Zealand, Norway, Poland, Portugal, Russia, South Africa, Spain, Sweden, Switzerland, Turkey, United Kingdom, United States of America, and Ukrai
6.  Davis, M. E.: The WTO Agreement on Basic Telecommunications Services, *Pacific Telecommunications Review*, Vol. 19, pp. 10-14, September 1997
7.  World Trade Organization: *The design and underlying principles of the GATS*, <http://www.wto.org/wto/services/services.htm>.
8.  Associate Administrator for Space Transportation: *ST News Bulletin*, Newsletter of Associate Administrator for Space Transportation, Volume IV, No. 4, page 2 February 8, 2000. "Thirty-six commercial launches were conducted in 1999 by launch providers in four countries and the international Sea Launch partnership. The United States captured 36 percent of the commercial launch services market with 13 launches. Russia also captured 36 percent of the market with 13 launches and Europe's Arianespace conducted eight commercial launches for a 22 percent share. Sea Launch and the China Great Wall Industry Corporation each had a three percent market share with one launch each"
9.  Isakowitz, S. J.: *International Reference Guide to Space Launch systems*, 2nd ed., updated by Jeff Samella, American Institute of Aeronautics and Astronautics, Washington, D.C., 1995 and cited in Elbert, B.R., *Introduction to Satellite Communication*, 2nd ed., Artech House Publishers, p. 405, 1999
10. Jakhu, R. and Tolyarenko, N: *Competition in World Space Launch Services?* International Space University, France, September 1998. "Competition is based on the assumption that: (i) market forces determine the price of a product, (b) there is free entry and exit from a market, (c) there is no direct government incentive or interference"

11. *Id.* "The trade in launch services is governed by bi-lateral agreements between the United States of America and China, Russia and Ukraine"... respectively. See also *infra* note 22

12. World Trade Organization: *Telecommunication Services, Background Note by the Secretariat,* S/C/W/74, December 8, 1998

13. Wassenaar Arrangement: *The Wassenaar Arrangement on Export Controls for Conventional Arms and Dual-Use Goods and Technology, Initial Elements,* Adopted by the Plenary of July 11-12, 1996 (hereinafter Initial Elements)

14. Wassenaar Arrangement: *Appendix 5 to the Initial Elements,* <http://www.wassenaar.org/list/Table%20of%20Contents%20%2099web.html#Dulist>.

15. Wassenaar Arrangement: *Appendix 5 to the Initial Elements,* <http://www.wassenaar.org/list/Table%20of%20Contents%20%2099web.html#ML>.

16. Stockholm International Peace Research Institute: *The Missile Technology Control Regime: An Information Paper,* <http://projects.sipri.se/expcon/mtcr_informationpaper.htm>.

17. The Annex contains (a) Category I items which include complete rocket systems (including ballistic missiles, space launch vehicles and sounding rockets) and unmanned air vehicle systems (including cruise missile systems, target and reconnaissance drones) with capabilities exceeding a 300km/500kg range/payload threshold; production facilities for such systems; and major sub-systems including rocket stages, re-entry vehicles, rocket engines, guidance systems and warhead mechanisms. (b) Category II items which include complete rocket systems (including ballistic missile systems, space launch vehicles and sounding rockets) and unmanned air vehicle systems (including cruise missile systems, target drones and reconnaissance drones) not covered in Item I, capable of a maximum range equal to, or greater than, 300km. Also included are a wide range of equipment, material and technologies, most of which have uses other than for missiles capable of delivering weapons of mass destruction

18. Arianespace: *Shareholders,* <http://www.arianespace.com/about_shareholders.html>. "France backs the ARIANESPACE Launcher with 57.06% shareholding of the 12 European countries participating in the Ariane program". See: Starsem: *The Companies that created Starsem,* <http://www.starsem.com/web_in/starsem.htm>. "France, through its Aerospatiale Matra company created a company responsible for marketing and providing associated launch services of the Soyuz Launchers. See also: Fenema, H, P van.: *supra* note 3 at 256 "French participation in the international trade of launch services is subject to the provisions of: The Convention for the establishment of the European Space Agency, 1975; and The Declaration by certain European governments relating to the Ariane launcher production phase, 1980"

19. Wassenaar Arrangement: *French Policy on Export Controls For Conventional Arms and Dual-Use Goods and Technologies,* <http://www.wassenaar.org/docs/fr1_eng.pdf>.

20. See: Stockholm International Peace Research Institute, *France,* <http://projects.sipri.se/expcon/natexpcon/France/france.htm>.

21.  See:   Stockholm   International   Peace   Research   Institute:   *Ukraine*,
     <http://projects.sipri.se/expcon/natexpcon/Ukraine/ukraine.htm>.
22.  Participation in the launch services industry by Ukraine has been conducted
     through a company (KB Yuzhnoye) providing the first two Zenit-3SL stages,
     launch vehicle integration support and mission operations with a 15%
     shareholding in the Seal Launch partnership. KB Yuzhnoye also participates in a
     consortium (ISC Komostras) that markets the DNEPR (a modified SS-18 ballistic
     missile). Participation is subject to the provisions of: The Agreement between the
     Government of the United States of America and the Government of Ukraine
     regarding international trade in commercial launch services, 1995; and The
     Satellite Technology Safeguards Agreement 1998. With the signing of a document
     between the United States of America and the Ukraine on June 5th 2000, the
     Agreement between the Government of the United States of America and the
     Government of Ukraine regarding international trade in commercial launch
     services has been suspended. The document lifts all restrictions on commercial
     launches of Ukrainian satellites and booster rockets
23.  Decree No. 117/98 of February 13, 1998 on State Export Controls in Ukraine
24.  Decree No. 422/99 of April 21, 1999
25.  Annex to the Statute on the Procedure for Controls on the Export, Import and
     Transit of Items, of Missile Technology, as Well as Equipment Materials and
     Technologies, That are Used in the Creation of Missile Weaponry approved by
     Decree No. 563 of the Cabinet of Ministers of Ukraine of July 27 1995
26.  Both authorities were established by Presidential Decree No. 1279/96 of 28
     December 1996 'On Further Improvements to the State Export Controls'
27.  Brisibe, T.C.: Policy and Regulatory Developments in Asia-Pacific after the
     GMPCS MoU and WTO General Agreement on Trade in Services: A Case for
     GMPCS System Operators, *Pacific Telecommunications Review*, Vol. 21, No. 3, First
     Quarter 2000
28.  *Id.* at page 31. "The levels include: 1) the main text containing general principles
     and obligations that all members have to apply on issues concerning, total
     coverage of internationally traded services (Article I), the Most favored Nation
     (M.F.N.) principle of non-discrimination (Article II), national treatment (Article II),
     transparency (Article III), regulations (Article VI), international payments (Article
     XI), individual countries' commitments (Part III, Articles XVI, XVII and XVIII) and
     progressive liberalization (Article XIX). 2) annexes, dealing with rules for specific
     sectors. and 3) Individual countries' schedules on specific commitments to provide
     access to their markets."
29.  This paper contains pro-competitive regulatory principles relating to competition
     safeguards,   interconnection,   universal   service,   transparency   of   licensing,
     independence of regulators and scarce resource allocation
30.  Fenema, H, P van.: *supra* note 3 at 322
31.  Johnson-Freese, J.: Life after the Cox Report: Technology Transfer and Export
     Controls, *Pacific Telecommunications Review*, Vol. 21, No. 1, Third Quarter 1999

# Report on Panel Discussion 5:

# What is the Appropriate Role of Governments in Future Space Transportation?

**L. Swift and C. Wagner,** International Space University, Strasbourg Central Campus, Parc d'Innovation, Boulevard Gonthier d'Andernach, 67400 Illkirch-Graffenstaden, France

e-mail: swift@mss.isunet.edu, wagner@isu.isunet.edu

**Panel Chair : F. Engström, ESA Headquarters, France**

**Panel Members:**

**T.C. Brisibe,** Le Goueff Avocats, Luxembourg
**A. Derechin,** RSC Energia, Russia
**P. Eymar,** Aérospatiale Espace et Défense, France
**E. Haase,** International Launch Services, USA
**C. Lovell,** United Space Alliance, USA
**K. Sazuki,** University of Sussex, UK
**J. Schnaars,** The Boeing Company, USA
**A. van Gaver,** Directorate of Launchers, ESA

Questions brought up during the discussion included:

- How can launch range safety be guaranteed?

- Who will accept the risk in the future?

- Is everybody competing under the same conditions?

- How would the market pay for new developments if the government doesnot?

- Will governments work together to develop the space transportation market?

The final question posed to the panel was "if you were a politician, what would be the most important decision you could make which would lead to lower launch costs and to a larger market?" The entire panel responded in turn, and the responses were as follows:

*M. Rycroft (ed.), The Space Transportation Market: Evolution or Revolution?*, 275–276.
© 2000 *Kluwer Academic Publishers. Printed in the Netherlands.*

**K. Suzuki** — create public demand, use simpler and off-the-shelf technology, and consider launchers as just one transportation device (like a truck)

**C. Lovell** — since the government's role is as a user, it needs to learn how to be a consumer of launch services, to put policies and regulations in place for public safety and then to "get out of the way"

**J. Schnaars** – need a major disaster — such as an asteroid approaching the Earth — or to find aliens in space!

**P. Eymar** — to open up new space frontiers, new space transportation is required, which requires government-funding

**E. Haase** — the government, as a user in the launcher market place, should be concerned with technology developments and with safety

**T. Brisibe** — the classical method is to allow a new market to develop, but would the market demand generate sufficient revenue? As there are so many variables, and several solutions, there is no one simple answer yet

**A. Derechin** — need to simplify of technology transfer procedures, and support the creation of real customers for new space launchers

**A. von Gaver** – try to have long term views. Past activities (such as Apollo) and the ISS were political acts; and need now to look at science (structure of the Universe, use Moon for science and resources) as the main justification for space programs, and commercial spin-off is secondary.

The conclusion (or consensus) seems to be that the proper role for government in space transportation is to fund the development of new technologies and assure public safety. It was also agreed that, with the present actual launch costs, is not possible to pay for new developments. The risks of space travel should be held by the companies involved so that the government doesnot have to pay for failures.

# Session 6

# Concluding Panel Discussion:
# Highlights and Lessons Learned in all Sessions

Session Chair:

**K. Doetsch,** International Space University, France

# Report on Panel Discussion 6:

# Concluding Panel Discussion:
# Highlights and Lessons Learned in all Sessions

**M. Breiling, H. Trinh,** International Space University, Strasbourg Central Campus, Parc d'Innovation, Boulevard Gonthier d'Andernach, 67400 Illkirch-Graffenstaden, France

e-mail: breiling@mss.isunet.edu, trinh@mss.isunet.edu

**Panel Chair : K. Doetsch,** International Space University, France

**Panel Members:**

**R. Akiba,** Space Activities Commission, Japan
**K. Bahrami,** MSS5 student
**J. Gantt,** Mizrack and Gantt, USA
**D. Heydon,** Arianespace, Inc., USA
**P. van Fenema,** International Institute of Air and Space Law, Leiden University, The Netherlands

The panel members were introduced by **K. Doetsch**; they are international leaders who focus on the future of space transportation from different perspectives. The session began with their individual reflections on, and summaries of, the Symposium.

**D. Heydon** gave his viewpoint, that of the commercial launch service provider. He stated that several changes in the industry need to take place in order for space transportation to rise from its infancy. Operators have consistent, but conflicting, desires. These include high reliability and performance, but not using new hardware. While satellite providers want high flexibility, they also want no delays and less risks. In the most extreme case, he stated that ELVs and RLVs cannot coexist, that evolved ELVs provide better value for money, and that RLVs will eventually dominate the market.

**P. Fenema** introduced the notion that the space launch industry may not be a "normal" industry, and that normal economic rules may not apply. Instead of comparing the launch industry with today's aviation standards, it should be compared with yesterday's aviation standards, with its poor reliability and inherent risk. With the terrestrial alternatives that telecommunications have, if space transport does not "shape up", it will lose its telecommunications clients.

*M. Rycroft (ed.), The Space Transportation Market: Evolution or Revolution?*, 279–280.

**R. Akiba** stated that governments should invest in the future by providing the infrastructure which the space launch market requires, and that the launchers should make suggestions to their customers. He agreed with the first two speakers that new technology for space transportation is a "chicken-and-egg" problem. Also, having easy access to space is the future duty of the space transportation industry.

**J. Gantt** summarized the space policies and laws that have been influential since the start of the space industry. He emphasized four key points:

- Need evolution, not revolution

- Be proactive, not reactive

- Do what's necessary, not what's just desirable

- Aim for simplicity and less risks.

Finally **K. Bahrami's** remarks opened up a more general discussion. Many interesting and provocative topics had been suggested for the MSS5 Team Project to consider in the coming three months. The general agreement is that there are constraints that prevent the maturation of the space industry. The governments, the launch providers, and the satellite builders must cooperate to mitigate the problems that we face today.

The final remarks of the Symposium were made by **Y. Fujimori**, introducing next year's Symposium title — "Smaller Satellites, Bigger Business: Concepts, Applications, and Markets for Micro- and Nano- Satellites in a New Information World."

# Poster Papers

# Aspire — An Independent British Launch Programme

**A. Baker**, AspireSpace, UK

e-mail: ADAM@ROCKETRY.FREESERVE.CO.UK

**Abstract**

This paper examines the position and potential of a small group of principally British engineering enthusiasts and entrepreneurs, AspireSpace, whose goal is to develop a privately financed nanosatellite launch service. AspireSpace has been developing small sounding rockets throughout the 1990s, and is now proficient with a number of the techniques required for a small, simple and low cost launch vehicle. To date, a number of attempts to develop small launchers have failed, but AspireSpace believes that it can be one of the first to succeed. The reasons are outlined in this paper, and include:

- A high reliability, simple, and safe propulsion unit, using a hybrid chemical rocket engine
- A novel funding mechanism, avoiding government funding but including sponsorship from industry and low-cost development in a university environment
- Integrated design, development and launch from a single site on the UK mainland.

## 1. The Mission of AspireSpace

AspireSpace formed in 1993, after a successful one year collaboration between four British Universities. A rocket vehicle (Aspire 1) was designed, built, and flown with on-board telemetry and a video camera payload, and subsequently recovered. Aspire 1 reached an apogee of 3.5km, believed to be a record for a UK amateur group at the time. On returning to the UK and evaluating the programme, team members perceived then, and still perceive now, an absence of government supported launch vehicle work in Britain. Furthermore, the general public in Britain are generally unaware of both the benefits of space and the level of space spending in its own country, and receive insufficient high quality hands-on training in British science and engineering degrees, despite considerable past British excellence in space endeavours. AspireSpace members therefore decided to build a non-government funded rocket which would reach space, a first in the UK and the world. The programme required to achieve this would act as a focal point for young people's space aspirations and as a means to educate and inspire them, through the wide range of interdependent systems issues required for a launch vehicle.

## 2. The AspireSpace Programme

Following on from Aspire 1, members of the newly formed AspireSpace conceived Aspire 2, a vehicle designed to reach 100km altitude, carrying up to 10kg payload, with a gross lift-off mass of no more than 200kg (permitting rapid

283

*M. Rycroft (ed.), The Space Transportation Market: Evolution or Revolution?*, 283–286.
© 2000 *Kluwer Academic Publishers. Printed in the Netherlands.*

deployment by a small team without heavy-lifting equipment). Aspire 2 was to be propelled by a hybrid rocket engine developed 'in-house', and would be recovered substantially intact after a flight to enable rapid refuelling and reuse. Hybrid rocket engines were selected for the core of the programme because of their operational safety, relative simplicity compared to higher performance bi-propellant engines, cleanliness and legality within the UK, where solid rocket motor construction requires extensive licensing by the HSE.

Three issues have modified AspireSpace's goals:

•   The realisation that the jump between Aspire 1 (3.5 km apogee) and Aspire 2 (100 km apogee) was too great to be made in a single step

•   Assessment of the late 1990s launch vehicle market indicated a large excess of launchers for payloads in the > 1000 kg market [Reference 1], no launch capacity for 'minisatellites' of 100-500 kg and little dedicated, operationally flexible launch capability for payloads < 100 kg [Reference 2]

•   Advances in microminiaturisation resulting in very small but functionally capable 'nanosatellites' weighing < 10 kg [References 3, 4].

In order to overcome the technological hurdle cited in (1.) above, a staged or evolutionary development programme resulted. The programme commenced in 1996 with flights of small rocket vehicles (< 10 kg) propelled by off-the-shelf solid rocket motors, which were designed to permit testing of subsystems for larger, trans-atmospheric vehicles and give AspireSpace credibility in the UK. Subsequently in 1998 small hybrid engines were developed, allowing team members to gain experience in the design, fabrication and testing of propulsion units. Combining rocket vehicles and propulsion systems of increasing complexity, overcoming technical and programmatic hurdles at each stage before proceeding is expected to lead to construction of an Aspire 2 scale vehicle within the decade. How this capability will develop from Aspire 2 and whether it can be converted into a commercial venture depends on the developing market for very nanosatellites and new sub-orbital applications over the next decade.

## 3.    The Commercial Marketplace

Despite the absence of detailed predictions regarding the size of the market for nanosatellites and their launcher services, it is felt that launchers for 'small' and 'very small' in addition to 'large (e.g. DBS and telecommunications) payloads are a potential rapid growth area over the next few years. Nanosatellites are already being demonstrated on-orbit, by microsatellite companies such as Surrey Satellites (SNAP-1, PICOSAT), the military (USAF TechSat-21) and by universities (Weber State University JAWSAT, Stanford University OPAL). Applications such as formation flying for synthetic aperture radar, MEMS microtechnology attitude control and on-board sensing using GPS, and constellation networking and software protocols have been demonstrated. Therefore, the development of an exclusive, inexpensive and operationally flexible launch service for micro- and nano-satellites is timely.

Such a service would need to be reliable; this would be enabled by the use of a simple vehicle maximising the employment of off-the-shelf components and powered by pressure-fed hybrid rocket engines, with the majority of development and integration done by a small independent team working at a low overhead cost. The launch service would also need to have a low total cost to enable competition with secondary payload launches such as Ariane V, with converted ICBMs such as Shtil and Start, and with 'mothership' launches such as from the Space Shuttle, the ISS or larger satellites. The lower overall cost would need to be emphasised in the face of the likely high total cost per kg delivered to LEO or polar orbit. Finally, the flexibility to launch with short notice, days or weeks instead of several months or even years as is currently the case, and with frequent launch opportunities, would appeal to a wide customer base.

Studies of historical small launch vehicles have indicated that securing payload customers in advance is an essential feature of development, if investor confidence is to be maintained. This was demonstrated with the Pegasus vehicle, where the first few launches were guaranteed by DARPA, in contrast with the problems which have beset numerous other launcher projects such as Orbital Express, PacAstro, Vega and more.

## 4.    Current Programme Status

AspireSpace has developed considerable expertise and credibility with small hybrid rocket engines in the last few years, and has demonstrated both a small N2O/HDPE engine which has potential for upper stage applications and a prototype LOX/HDPE engine sized for boosting payloads on sub-orbital trajectories. In addition to the propulsion system, An gyroscopically stabilised rocket vehicle employing gimbal mounting of the rocket motor  has been successfully flown, fluid and ablation cooled rocket nozzles have been tested, and a remotely steerable recovery system together a radio transmitter package capable of transmitting full colour video and telemetry are all under development at present. These research areas and more have been suggested as university student final year projects, with interested students first being introduced to AspireSpace through a National Rocket Championship organised by AspireSpace in conjunction with the UK Rocketry Association and the UK Students for the Exploration and Development of Space.

AspireSpace continues to demonstrate subsystems on small rocket vehicles weighing ~ 10 kg and at altitudes up to 6 km. Where propulsion was initially by commercial off-the-shelf solid rocket motors, AspireSpace now uses hybrid engines for increased safety, controllability and performance and is preparing to launch its first hybrid powered test vehicle. An intermediate sub-orbital vehicle designed around the current prototype LOX/HDPE engine is the next stage of the programme, and is intended to be funded by a revenue stream acquired from a novel application of hybrid rocket engines. Once the sub-orbital vehicle has demonstrated the technology synergy required, at an acceptable cost, development of the dedicated nanosatellite launch vehicle will be begun in earnest.

The good relationship between AspireSpace and UK licensing authorities suggests that polar and Sun-synchronous launches from the UK mainland will be achievable. In the event of a market demand for equatorial or inclined orbit insertion, use of air launch to increase operational flexibility is an option and has already been shown to be technically viable for a small (< 10kg) payload. UK customers for the vehicle are already being scoped, in the form of small businesses expanding into the space market. The potential for linking the currently diversifying UK internet technology market with a high profile space application such as rocket vehicles and satellite launching is being explored. Furthermore, a number of other rocket development activities are occurring in the non-governmental sector, and there is a strong possibility of AspireSpace linking with one of more partners to increase the rate of technology development, further enhancing the UK position as a principal provider of launch services for the nanosatellite market of the new Millennium.

### References

1.   Caceres, M. A.: Industry Faces Launcher Excess, *Aviation Week & Space Technology*, pp. 135-136, January 17 2000.
2.   Launch Vehicles, *Aviation Week & Space Technology*, pp. 140-149, January 17, 2000.
3.   The Aerospace Corporation: *Picosatellites launched aboard converted Minuteman II booster* <http://www.aero.org/news/current/picosat-00.html>. May 1, 2000.
4.   SSTL: *Nanosatellites - SNAP-1*, <http://www.sstl.co.uk/services/mn_nanosatellites.html>. May 1, 2000.

# Launching Small Spacecraft —
# The Surrey Space Centre Experience

**J. Keravala**, Surrey Satellite Technology Ltd, Surrey Space Centre, UK

e-mail: j.keravala@ee.surrey.ac.uk

**V. Lappas**, Surrey Space Centre, University of Surrey, UK

e-mail: v.lappas@ee.surrey.ac.uk

**Abstract**
The development of 'smaller, faster, cheaper and better' spacecraft has enabled many countries and universities to build, launch and operate their own small satellite thereby providing them with direct access to the advantages of space. Surrey Satellite Technology Ltd (SSTL), based at the Surrey Space Centre at the University of Surrey, has built and launched 15 multi-mission micro- and mini-satellites into low Earth orbit. Its successful approach of 'Affordable Access to Space' has enabled SSTL to develop unique experience of launching microsatellites in a cost effective manner. This paper will outline the SSTL experience in launching small spacecraft and also discusses issues related to the future of the small satellite and constellation commercial launch market.

## 1. Introduction to Surrey Activities

Traditionally, space missions have been large, expensive programmes taking many years and limited to relatively large and wealthy nations, international agencies or large commercial programs. However, this is no longer the only path into space: advances in microelectronics have made small-scale space missions very affordable, while also delivering impressive and valuable results. The development of such intelligent, low-cost, long-life, small satellites has been pioneered at Surrey Satellite Technology Ltd (SSTL), based at the Surrey Space Centre located at the University of Surrey in the UK. Over the last twenty years, Surrey has built and launched 15 multi-mission micro/mini-satellites into low earth orbit.

Three families of spacecraft are developed at Surrey ranging from sophisticated 5 kg nanosatellites through to microsatellites and including the larger 350 kg microsatellites, of which UoSAT-12 was the first example. All these spacecraft utilize innovative, yet cost effective technologies and fast turnaround program management. Each phase of the Surrey microsatellite development cycle from design, manufacture, testing and operations, serves to reduce mission cost and turn-around time. As part of each mission, increased effort is placed on the procurement, management and implementation of the launch of Surrey's spacecraft, listed in Table 1.

*M. Rycroft (ed.), The Space Transportation Market: Evolution or Revolution?*, 287–289.
© 2000 *Kluwer Academic Publishers. Printed in the Netherlands.*

| Microsatellite | Launch | Orbit | Customer | Payloads |
|---|---|---|---|---|
| UoSAT-1 | 1984-D | 560 km | UoS | Research |
| UoSAT-2 | 1984-D | 700 km | UoS | S&F, EO, rad |
| UoSAT-3 | 1990-A | 900 km | UoS | S&F |
| UoSAT-4 | 1990-A | 900 km | UoS/ESA | Technology |
| UoSAT-5 | 1991-A | 900 km | SatelLife | S&F,EO, rad |
| S80/T | 1992-A | 1330 km | CNES | LEO comms |
| KitSat-1 | 1992-A | 1330 km | Korea | S&F,EO, rad |
| KitSat-2 | 1993-A | 900 km | Korea | S&F,EO, rad |
| PoSAT-1 | 1993-A | 900 km | Portugal | S&F,EO, rad |
| HealthSat-2 | 1993-A | 900 km | SatelLife | S&F |
| Cerise | 1995-A | 735 km | CNES | Military |
| FASat-Alfa | 1995-T | 873 km | Chile | S&F,EO |
| FASat-Bravo | 1998-Z | 835 km | Chile | S&F,EO |
| Thai-Phutt | 1998-Z | 835 km | Thailand | S&F,EO |
| UoSAT-12 | 1999-S | 650 km | SSTL & Singapore | EO, Comms |
| Clementine | 1999-A | 735 km | CNES | Military |
| SNAP-1 | 2000-C | 650 km | SSTL | Technology |
| Tsinghua-1 | 2000-C | 750 km | PR China | EO, Comms |
| TiungSAT-1 | 2000-S | 1020 km | Malaysia | EO, Comms |
| PicoSAT | 2001-? | 650 km | USAF | Military |

**Table 1.** University of Surrey microsatellite missions

One of the most important obstacles which makes access to space very expensive is launching spacecraft to space. A sustained, commercial microsatellite programme requires predictable and regular access to orbit through formal launch service contracts — since the ability to construct sophisticated yet inexpensive microsatellites must also be matched by a correspondingly inexpensive method of launch into orbit.

## 2.    Details of Surrey Activities

The establishment of launch services for small satellites presents a set of unique challenges. The opportunity for rapid spacecraft turn-around requires the capability to secure the appropriate launch opportunity at an affordable cost, often on short timescales. Most small spacecraft are secondary payloads launched into orbits with their primary payloads. The ability to secure such payloads calls for flexibility with launch schedules and commencement of mission operations. Despite these potential limitations, such launch opportunities provide highly economical options. The earliest Surrey spacecraft were first launched on the US Delta. In 1988, Arianespace developed the Ariane structure for auxiliary payloads (ASAP) specifically to provide, for the first time, regular and affordable launch opportunities for 50 kg microsatellites into both LEO and GTO on a commercial basis. Whilst the ASAP has been key to catalysing the development of microsatellites, it cannot now provide the number of launch opportunities needed to meet the burgeoning growth of small satellites. Recently, the most cost effective options for launching small spacecraft are found using Russian and Ukrainian launch vehicles. In addition to existing launch vehicles that have been in service for a many years, large stockpiles of ICBMs in the CIS have become available for use as small launchers through the demilitarilisation programme (e.g., SS-18/Dnepr, SS-19/Rockot, SS-25/START). SSTL co-operated with ISC Kosmotras (Moscow) to convert the SS18 ICBM into the first Dnepr small satellite launcher for the successful launch of the UoSAT-12 minisatellite from a silo at Baikonur in April 1999.

A successful launch program for a small satellite involves identification of available launch options, contractual arrangement and procurement, spacecraft interface development, mission analysis, technical management, transportation, international licensing and insurance, including attention to numerous logistical issues. Such aspects are to be considered in all spacecraft missions, though their relative influence on the time-scales of smaller satellite missions can be greater. Surrey has worked successfully with numerous agencies, and ensured that the management of launch campaigns has maintained the same innovative philosophies as adopted during spacecraft development. Many new small launchers currently available or being developed (Pegasus, Athena, VLS, Vega) also offer opportunities for small spacecraft in secondary or primary configurations, though some are expensive when compared to the correspondingly small satellite cost. The same radical approach is required for launchers as has been pioneered for small satellites, ultimately reducing the cost of launch vehicles significantly.

With the increasing technical capabilities of small spacecraft, the range of applications is increasing. A number of new small spacecraft constellations will be developed at Surrey. Microsatellites provide affordable options to develop a global constellation of satellites for earth observation, disaster monitoring or communications. Over the next few years, several constellations consisting of a total of 25 small spacecraft will be designed and built at Surrey, in addition the continuing single satellite programs. Over the coming years, Surrey will have a requirement to launch tens of spacecraft. This makes the need for frequent and inexpensive launch options more important than ever.

# Telemetry and Tracking System for the H-IIA Launch Vehicle

**N. Kohtake, H. Kawabata, K. Teraoka, M.Katahira, H. Takatsuka,** H-IIA Project Team, National Space Development Agency of Japan, World Trade Center Bldg., 2-4-1, Hamamatsu-cho, Minato-ku, Tokyo 105-8060, Japan

e-mail: kohtake.naohiko@nasda.go.jp

**Abstract**
The National Space Development Agency of Japan (NASDA) has been developing the H-IIA launch vehicle, a new two-stage expendable launch vehicle to meet the changing launch demands in the 21st century, with lower costs and higher reliability. Regarding the telemetry and tracking system, three new subsystems have been developed based on our experience with H-II. This paper describes the three new subsystems; a MPEG-I Onboard Monitoring System, a GPS Receiver System for Flight Safety, and a High-rated Data Acquisition System.

## 1. Introduction

The telemetry and tracking system of the H-IIA has been developed to make the launch vehicle more cost effective and to meet many diverse demands [Reference 1]. The system enables the H-IIA manufacturing cost to be 35 % lower than the H-II cost by simplifying the integration of components and interfaces, adopting hybrid ICs, G/As, FPGAs, and implementing Built In Test functions. In order to meet different demands, three new subsystems have been developed as discussed in Section 2.

## 2. Three New Subsystems

### 2.1 MPEG-I Onboard Monitoring System

The MPEG-I onboard monitoring system for observation of the flight sequence is composed of CCD cameras, lamps and a picture compression unit (PCU) including a real-time encoding/decoding (CODEC) LSI in MPEG-I format [References 2,3]. The system is manufactured by refining a commercial digital video camera system for space-use; this is cheaper than creating a dedicated product for use in space. The main benefit of using this system is to acquire visual data of some important events, and to evaluate flight status more precisely than using numerical data. In addition, the visual data can provide H-IIA flight information to the general public. During the first test flight, satellite separation will be observed. Fig. 1 shows all points to be monitored. The system can also monitor propellant motion inside the liquid hydrogen (LH2) tank in the coasting phase.

*M. Rycroft (ed.), The Space Transportation Market: Evolution or Revolution?*, 291–293.
© 2000 *Kluwer Academic Publishers. Printed in the Netherlands.*

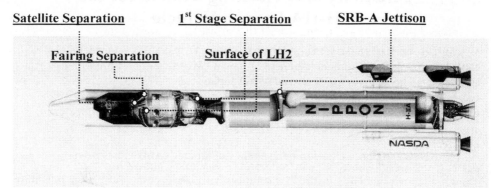

**Figure 1.** Monitoring points for the first test flight of H-IIA

## 2.2   GPS Receiver System for Flight Safety

In addition to the radar transponder which is mainly used for tracking, a GPS receiver has been developed based on the past experiments that NASDA had performed [References 4, 5, 6]. It is expected that the GPS receiver will replace the radar system for flight safety and that position and velocity data from the GPS receiver will be used along with those of the inertial measurement unit to improve precision. An engineering model will be mounted on the first test flight. Fig. 2 shows the configuration of the GPS Receiver.

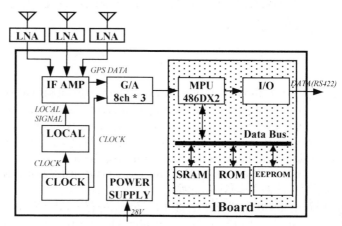

**Figure 2.** Diagram of GPS receiver

*2.3   High-rated Data Acquisition System*

The data which are gathered by the propulsion, engine and other systems are transmitted to the ground stations by telemetry transmitters for technical feedback during/after the flight. However, these data are not enough to evaluate a flight. To solve the problem of the lack of data, optional data acquisition units for other parameters will be mounted. With this new subsystem, data on acoustics, vibration, acceleration, etc., will be gathered at a maximum sampling rate of approx. 10000 times/sec and will be transmitted. In addition, as a service for users, data on the payload environments and on the payload itself will also be transmitted using this new subsystem.

## 3.   Future Work

NASDA currently manages some downrange stations to monitor the flight conditions of launch vehicles using telemetry. In the future, we are planning to develop a telemetry system through a data relay satellite to meet more different launch demands. We consider that our SELENE (Selenological and Engineering Explorer) mission in 2004 will be the first opportunity for using this system.

## 4.   Conclusion

This paper has presented three new subsystems of the H-IIA telemetry and tracking system which will contribute to meet various launch demands for the H-IIA users. Now, we are testing them in detail with the H-IIA ground test vehicle. The first test flight is scheduled to be launched early in 2001 from the Tanegashima Space Center.

**References**
1.   Watanabe, A., et al.: *H2-H2A Redesign for More Efficient and Active Space Development*, IAF-98-IAA.1.1.01, 49th International Astronautical Congress, 1998
2.   Kurashige, T., et al.: *Development of MPEG Camera*, pp. 31-35. IEEE ISCE'97, 1997
3.   Kohtake, N., et al.: *Onboard Monitoring System to Observe the Flight Sequence of H-IIA Launch Vehicle*, 51st International Astronautical Congress, 2000
4.   Kawano, I., et al.: *Relative GPS Navigation for an Automated Rendezvous Docking Test Satellite ETS-VII*, ION-GPS '97, 1997
5.   Suzuki, H.: *Application of GPS to Spacecraft in NASDA*, pp. 1-8. SANE98-117 IEICE, 1999
6.   Kohtake, N., et al.: *Status of Developing GPS Receiver for H-IIA Launch Vehicle*, pp. 63-69. SANE98-117 IEICE, 1999

# The Space Launch Services Industry: Indicators and Trends

C. Mace, C. Christensen, G. Lucas, Futron Corp., 7315 Wisconsin Ave, Ste. 900W, Bethesda, MD 20814, USA

e-mail: cmace@futron.com, cchriste@futron.com, glucas@futron.com

**Abstract**

This paper provides a qualitative and quantitative analysis of the international space launch services industry with particular emphasis on forecasts of future demand and supply for space transportation services through the year 2010. Futron will provide numerical indicators as well as analyses to demonstrate the drivers and effects associated with the launch services industry. Futron Corporation has recently completed a number of large research products on the underlying demand for satellite and launch vehicle services. A detailed look into some of Futron's findings about the key drivers for growth in launch vehicle services and about possible future oversupply, are given.

## 1. Launch Vehicle Selection Factors

The commercial launch market has grown more significant and more competitive in recent years. In 1997, for the first time, revenue from commercial launches exceeded the revenue from non-commercial launches worldwide. In the next five years, 12 new vehicles are expected to enter the market to compete for commercial launches. As a result, launch service providers have had to modify their marketing strategies to capture a share in the larger commercial market and remain competitive with the growing number of service providers. These new marketing strategies implemented by the space industry are based on the selection factors that influence the procurement by the satellite operator.

The most influential selection factors in the commercial launch market, after basic lift capability to orbit is considered, are the financial characteristics of the service, which include price and terms and conditions of the contract. The entry of new vehicles from the former Soviet Union and China have added pressure to launch service providers to price launches at competitive rates. One instrument used by launch service providers to reduce prices is the "bulk" sale of launches. Launch service providers are also offering more favorable terms and conditions to satellite operators and manufacturers. Insurance and equity investment in satellite systems are examples of the terms which service providers offer in a contract. Other important selection factors are the technical characteristics of the vehicle. Reliability of the vehicle, fairing size, orbital insertion capability, and ability to dispense multiple satellites from the same vehicle are all characteristics that influence the selection of a vehicle.

*M. Rycroft (ed.), The Space Transportation Market: Evolution or Revolution?, 295–296.*
© 2000 *Kluwer Academic Publishers. Printed in the Netherlands.*

## 2.    Launch Vehicle Demand

Futron projects that on-orbit demand for satellites will translate into an average of about 20 commercial GEO satellites launched each year, with significant year to year variations, over the next ten years. Demand for GEO launches is relatively flat despite a significant growth in demand for satellite services. This disparity is due to the increasing capability of GEO satellites and the resulting reduction of the number of satellites needed to provide a given level of service. New technological developments in satellite manufacturing are enabling the bus size and transponder complement of GEO satellites to increase substantially. Larger bus size allows more stationkeeping propellant to be stored aboard a satellite, yielding a longer service life. Additional transponders provide more communication capacity, also increasing the ability of a satellite to deliver a given level of service, and data compression techniques are being introduced.

In the non-geostationary orbit (NGSO) launch market, the projected demand for launches has significantly decreased from last year. With the failure of Iridium and the financial difficulties of ICO, several NGSO systems have either scaled down their initial system numbers or cancelled plans for a system. Further, data service providers are focusing on the use of GEO satellites for infrastructure to enter the market faster and at lower costs.

## 3.    Launch Vehicle Supply

Futron projects that there will be little growth in the number of commercial satellite launches over the next ten years. Demand will tend to shift to medium and heavy launch vehicles that are able to accommodate larger payloads. Competition is expected to increase, caused by the fact that global launch capacity will significantly exceed demand. At least 12 new expendable launch vehicles may offer a range of technical capabilities in the next five years; among these are Russian launch vehicles which should deliver launch services at very competitive prices.

Six American reusable launch vehicles are in development to compete with expendable rockets for satellite launches: Kistler Aerospace's K-1, Pioneer Rocketplane's Pathfinder, Rotary Rocket's Roton C-9, Space Access's SA-1, Vela Space Technology's Space Cruiser, and Lockheed Martin's Venturestar. Another vehicle, Great Britain's Spacecab/Spacebus, is in development by Bristol Spaceplanes Limited. The uncertainty of the NGSO launch market and the technical challenges of developing RLVs appear to have resulted in financing problems for all RLV developers.

# Investment Appraisal of Next Generation Launch Vehicles using Real Option Theory

**D. Perigo, M.R. Ayre,** College of Aeronautics, Cranfield University, Cranfield, Bedford, MK43 OAL, UK

email: d.perigo@cranfield.ac.uk, m.r.ayre@cranfield.ac.uk

### Abstract

The path to appraising an investment opportunity, and hence deciding whether to proceed and under what financial structure, is one of great importance to all commercial organisations. This is particularly true in the space transportation industry, where commercialisation has in recent years created an international launch market. At the present time, several organisations are involved in projects developing Next Generation Launch Vehicle (NGLV) technology. These projects are both extremely costly and present a considerable amount of financial risk; hence determining the optimal way in which to finance NGLV projects is a formidable problem. This paper applies the new discipline of Real Options to the appraisal of NGLV investments. This theory adds to the neo-classical DCF theory of investment evaluation through the consideration of an organisation's ability to make decisions based upon contingent information as and when it becomes available.

From the analysis undertaken it is shown that NLGV development is financially feasible and, more importantly, is feasible in a purely commercial context. It is also demonstrated that such a project need not necessarily be the most conservative option and that the scope of current financing can be considerably widened to provide greater flexibility and greater security for both the project and its investors alike.

## 1.  Consideration of Future Markets

Prior to the development of a next generation launch vehicle (NGLV), it is essential to have an indication of the potential customer base that this vehicle will serve and also the money available to those potential customers in order to buy the service that will be supplied. It is also essential to possess an accurate picture of the product range of the potential customer and the needs that such products will impose on any new vehicle to be designed. Further, insight must be gained into the future products of those customers and the proposed vehicle must be designed with such products in mind as well. Finally, a potential producer of a NGLV must, to some extent, examine the long-term viability of the customer markets before embarking on what is essentially a long-term investment plan. This is essential as, even with the downward pressure on project lead times within the aerospace industry, the development of a new launch vehicle is a far from a trivial task. With this in mind our first step in assessing the potential profit to be made from developing a next generation launch vehicle is to attempt to quantify the current and future market for launch services within the space industry as a whole.

*M. Rycroft (ed.), The Space Transportation Market: Evolution or Revolution?, 297–310.*
© *2000 Kluwer Academic Publishers. Printed in the Netherlands.*

For this analysis the launch market was broken down into the following sectors:

- Communications

- Earth Observation

- Science Missions

- Military

- Space Tourism.

The following sections summarise the conclusions of the initial analysis of these areas. The focus is, in the main, on the communications market as this is currently the only truly viable commercial sector and any commercial launch venture will need to match accurately the vehicle capabilities with the requirements of this customer base.

## 1.1 Communication

The current global telecommunications market as a whole has an estimated worth of $ 750 billion [Reference 1] and by 2005 should reach around $1200 billion. This covers all types of communications (voice, paging, broadband internet, television, etc.). Of the current market, around $ 30 billion is accounted for by the global satellite communications industry. This is by far the largest and most dominant customer group for the current launch services market and is expected to continue to be so for some time to come. It is hence the sector of the market that a launch system must be primarily designed to service.

The major segment, and major competitor, to the satellite communications industry is the terrestrial telecommunications industry. This is currently dominated by big corporations such as AT&T and MCI Worldcom but also populated by a large range of highly capital intensive start-ups (Qwest, Level 3 and Nextlink). This sector of the market has ambitious plans to install hundreds of thousands of kilometres of cabling, not only locally within large cities but between cities on a continental basis and also between continents. Their aim is to provide communications at all bandwidth levels, and do so on a very short time-scale (in the next decade). The driving force behind this colossal capital investment is the burgeoning information market, and the perception by business that connection to the e-world is essential. This shows no signs of abating and Qwest [Reference 2] forecast no surplus broad-band capacity for the foreseeable future. It is thus possible to say, with some confidence, that the projection for the broad-band satellite communications market (Internet, Multicast, etc.), at $13 billion by 2005 [Reference 3] is valid, and indeed may be exceeded by some margin as the full picture for broadband demand becomes apparent.

Significant growth is also foreseen in other sectors including Direct Broadcast Systems (DBS) for television (an increase of $ 14 billion). This is a sector where satellites can provide an instant solution, eliminating the need for ground infrastructure by broadcasting Direct to Home (DTH). The success and potential for this sector has been demonstrated already by numerous examples including the Astra and EUTELSAT platforms in Europe and Asiasat and Indostar in the East.

Growth in the Personal Communications (PC) sector is also forecast and, though the failure of Iridium in this arena has raised some questions, it seems generally accepted that this sector is still capable of generating significant revenues.

The benefits of using satellites for broadband communications are global coverage, with no need for cabling, and also bandwidth on demand, with the removal of the need to be cabled at the users' end. This allows the end user to pay for connection time rather than for the physical connection. Opportunities can also be seen in servicing the needs of developing countries and remote communities where cabling will become an expensive option. On the negative side there are problems with routine maintenance, and upgrades are intrinsically more difficult for satellite based systems. A more pressing problem, however, is the allocation of bandwidth and the lack of coherent international planning in this area.

The majority of the new systems for broadband Internet and Personal Communications are MEO/LEO constellations, typically around the 1.0 tonne mark. In contrast DBS satellites destined for GEO are considerably larger. As an example Astra's latest satellite weighs approximately 4.5 tonnes which may cause some problems for launch service providers as, for instance, the capability of Ariane 5 to GTO is only 6 tonnes. In this case a dual launch with a similar payload would be impossible.

In conclusion, the response of the global satellite communications industry to the opportunity provided by the huge growth in the information market has been swift. As a result, a significant and stable portion of the telecommunications market should be secured for many years to come. This swift response has resulted in a demand to launch a significant number of satellites, around 200 in the first wave with a follow on schedule of around another 500, into LEO and MEO slots. These satellites are all less than 2.5 tonnes in mass, with lifespans typically in the 7-15 year range. They thus provide a continuing market in upgrade and maintenance launches in the long term.

## 1.2   Earth Observation

Earth observations satellite launches have some potential as a secondary market from which extra revenue may be earned. From an examination of the current status of this sector it is suggested that the commercial viability of such systems, in the current market place, is questionable. This appears to be the case even if the operators take full advantage of the potential for future end products, e.g. close association with financial institutions for the prediction of crop futures, as well as the current uses for treaty monitoring / enforcement and environmental monitoring. In short government support has played and still does play a major role in this sector.   Work by Seynat et. al. [Reference 4] has shown that for commercial EOS, using a small constellation, significant reductions in the costs of spacecraft manufacture and launch are required before a reasonable return on investment is seen. The first of these concerns may be close to being addressed with the new production techniques being developed for communications constellations. The second is one of the prime reasons for developing a NGLV. Thus, while this market is currently small, a reduction in launch costs could open up this sector and provide additional launch revenue.

As a caveat to the above it is foreseen that this sector may come under pressure from emerging technologies such as High Altitude Long Endurance (HALE) Platforms and High Altitude Airships. Although these systems demonstrate smaller ground footprints they show greater flexibility with respect to servicing and re-configuration.  Hence reliance on an increase in this class of payload is deemed unwise.

## 1.3   Science Missions

No reason can be foreseen for the quantity of science missions not to continue at their current level. However, two classes of mission can be seen developing, namely big science and small science.

Big science missions are characterised by very specific launch requirements and are typically unique high profile missions for national agencies and hence, effectively outside the remit of a commercial launch vehicle. Small science missions, on the other hand, are assumed to be low mass piggy-back customers. It is thus assumed that competitive advantage for a launch vehicle will not be gained by catering for these missions.

## 1.4  Military

The servicing of the military launch market is considered beyond the scope of this paper as the aim of the paper is to address the commercial aspects of launch service provision. Military customers are ignored as a result of the problems in forecasting the regularity of military payloads, and of assessing the likelihood of being approached to launch such payloads when the need arises. There can be little doubt, however, that the military would be keen to take advantage of lower launch costs and any increase in the level of reliability seen in an NGLV program. The influential factor here would be the lack of control over the launch service provider and the international nature of the investor base. A solution to this may be to incorporate the needs of the military market with a technology leasing agreement. In this case the full launch system would be leased to, and operated by, the government in question.

## 1.5  Space Tourism

It can be said that the first space tourists have almost arrived with the foundation of Mircorp which aims to market the Mir space station facility commercially for multi-million dollar price tags . This can be assumed to represent the highest real term price that the space tourism market will see. This type of venture is the perceived starting point for the space tourism market. The $ 9 million recently asked by the Russians for a flight to Mir is, however, a price that few can afford and this will at best produce a small market serving a small number of "super rich thrill seekers". Several studies [References 5,6,7] have envisaged a larger market and even a mass market, but these, are dependent on a significant reduction in launch costs. This is an aim consistent with that of many of the potential new entrants to the launch market — Kistler, Kelly, Pioneer, Space Access and Rotary Rocket, to name but a few. In fact the potential new market for space tourism demonstrates significant goal congruence with the requirements for NGLV's in terms of reliability, abort capability and operational costs.

## 1.6  Implications for the Launch Vehicle Industry

Current forecasts [Reference 3] predict that the satellite communications industry will grow from its current $ 30 billion to $ 75 billion by the year 2005. This should correspond to an average of around 100 satellite launches a year. At today's prices, using current launchers, the available revenue for launching satellites can be estimated at around $ 10 billion, to be divided up amongst the entire launch industry. To succeed in this market launch service providers will have to demonstrate not only lower costs but also a high reliability and timeliness of launches. These factors, as much as any, will influence the choice of provider by the communications industry.

Currently established launch services will be stretched to service this requirement even with the capacity for multiple launches currently available. There is therefore an opportunity for new entrants to undercut traditional launch prices and carve out a niche market in the mass range that encompasses the majority of constellation satellites.

As a consequence of the above it is possible that downward pressure on the cost of launches will effectively reduce the available revenue that could be generated by the current world launch demand. However, these reductions in cost should also increase the feasibility of more marginal projects and hence increase the number of required launches.

This section has briefly touched on the potential future market for space tourism. Though not proven or, for that matter, popular with financial analysts, the potential size of this market is forecast to grow significantly with a reduction of launch costs. It is on this issue that the rest of the paper will concentrate, with specific attention being paid to the option for launch service providers to diversify into this market.

## 2.   Introduction to Real Option Theory

Traditional investment evaluation techniques (still the predominant method used today in the business environment) centre on the use of Discounted Cash Flow (DCF) analysis, also known as Net Present Value (NPV) analysis. In this case, the investment project is necessarily viewed as being static, i.e. the operating decisions are fixed in advance and, as such, give rise to the base set of incremental cash flows that are used to value the project. These cash flows are then discounted to take into account the time value of money to yield the Net Present Value, given by the equation [Reference 8]:

$$NPV = C_0 + \frac{C_1}{1+r} + \frac{C_2}{(1+r_2)^2} + \frac{C_3}{(1+r_3)^3} + K \frac{C_n}{(1+r_n)^n} \qquad (1)$$

where $C_0$ is the initial investment (i.e. a negative term in equation 1), higher $C$ terms represent the cash flows per time period and $r$ is the interest rate for equivalent-risk investments. The investment is then accepted or rejected depending on whether the NPV is positive or negative.

Because the view of the investment is a static one, no consideration is given to the role which management has to play. This has led many to now view NPV analysis as incomplete with regards to managed 'real-world' projects. In reality, an investment can be managed to maintain flexibility in as many ways as possible; this leads to the maintenance of options that give upside potential whilst protecting against loss. Therefore the ability to make decisions based upon options embedded in an investment can add significant increments of value to the investment. Before Real Option theory [Reference 9] was introduced, there was no way of quantitatively measuring the value of these management options, and as a consequence they have been largely ignored in investment appraisal.

*2.1  Application of Stock Option Theory to Investment Appraisal*

In the financial markets, stock options are used to manage risk in a way that is analogous to the management of risk through effective management of real options (the decisions to expand, enter into a new market, upgrade production, etc.). Therefore, standard stock option theory has been adapted to 'real' world investments: this allows a value to be attached to the options in the investment, thereby allowing the true value of the investment to be appreciated. In this instance, the NPV of the investment becomes:

$$NPV = C_0 + \sum \frac{C_t}{(1+r_t)^t} + PV_{opt1} + PV_{opt2} + ... \tag{2}$$

where $PV_{opt}$ is the present value of the options embedded in the investment, given by:

$$PV_{opt} = (N(d_1) \times P) - (N(d_2) \times PV(ex)) \tag{3}$$

where

$$d_1 = \frac{\log\left(P/PV(ex)\right)}{\sigma\sqrt{t}} + \frac{\sigma\sqrt{t}}{2} \tag{4}$$

$$d_2 = d_1 - \sigma\sqrt{t} \tag{5}$$

and

$N(d) = $ cumulative normal probability density function

$ex = $ exercise price of the option

$PV(ex) = $ present value of the exercise price

$t = $ number of time periods to exercise date

$P = $ price of the stock now

$\sigma = $ standard deviation per period of rate of return on stock

The ramifications of Real Option theory applied to investment appraisal are obviously important: as can be seen from equation (2), investments given a negative NPV from traditional analysis may well be found to have a positive NPV once options are factored in. Real Option analysis [Reference 9] is vital when considering highly unpredictable markets where many options exist, the future launch market being a perfect case-in-point.

## 2.2   Estimated Revenues from Space Tourism

In order to attempt to estimate the potential market for space tourism, several factors need to be considered. First and foremost is there public interest in such a service: how many people are interested in a trip to space? What is the price sensitivity of such a market? The potential revenue garnered from the space tourism is dependent on the price at which the service is offered. The development of the space tourism market is likely to undergo several phases, in a manner analogous to the development of other types of tourism, for example air travel. To begin with, air travel was the province of the very rich 'leisure class'. As the service matured and prices began to drop, more and more customers appeared, with the end result being a mass market (at least with regard to the first world).

A limited number of studies of the potential market have been conducted [References 5,6,7], and the results have been combined to produce the demand curve for space tourism in Fig. 1 and Table 1.

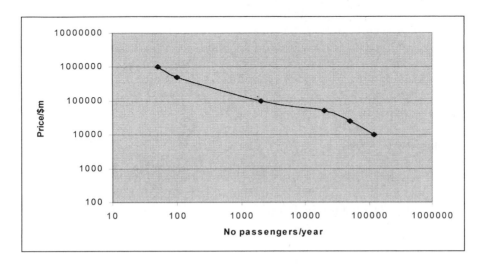

**Figure 1.** Estimated demand curve for space tourism services

| Customer Group | Number of passengers | Price/$ | Revenue/$m |
|---|---|---|---|
| 1 | 50 | 1.000,000 | 50 |
| 2 | 100 | 500,000 | 50 |
| 3 | 2,000 | 100,000 | 200 |
| 4 | 20,000 | 50,000 | 1000 |
| 5 | 50,000 | 25,000 | 1250 |
| 6 | 120,000 | 10,000 | 1200 |

**Table 1.** Estimated demand and revenue for Space Tourism as a function of price

From the surveys conducted it is therefore readily apparent that the market for space tourism is potentially huge. This becomes easier to accept when the fact that travel and tourism are now recognised as the world's largest industry with a turnover of $ 3.4 trillion [Reference 11]. The continuing fashion for travel, accompanied by increasing wealth creation and leisure time in the top brackets of society, will continue to fuel this industry into the 21st century. In the search for a novel and 'authentic' experience, a trip to space holds great appeal to a large section of the first-world population.

However, the figures quoted are heavily dependent upon many factors, which need to be addressed before space tourism can function effectively as an industry. Not least of these is the subject of safety. The quoted figures for mass-market space tourism are based upon the assumption that the experience will be at least *as* safe as air-travel is today. Whilst more risk would perhaps be acceptable to certain fringe demographics at the beginning of the industry, a mass market in space tourism would necessitate a level of safety and legislation commensurate with other forms of travel.

## 3.     Baseline Vehicle

Our baseline vehicle to satisfy this broad range of demands was selected to be an amalgam of the projected launch and operational costs given by many of the potential entrants into the launch vehicle market. The generic launch system can be summarised as follows:

•     Three operational launchers each comprising of a winged sub-orbital vehicle with an expendable kick stage for orbital injection

•     A payload capacity of 3000 kg to LEO at $ 4000 per kg.

•     Aircraft like ground handling with non-cryogenic fuels

•     A $ 400 million USD development cost spread over the first three years of the program

•     Start of operations in 2003

•     Operating costs of $ 100 million USD per year

•     A recurring launch cost of $ 1 million USD per launch.

In addition to the above, the following assumptions were also made:

•     A rival company offering a similar service was assumed as a competitor

• The capture of around a third of the market in satellite launches for LEO constellations based on [Reference 11] was assumed.

A simple financial model was constructed to assess the effect of including options, as well as of allowing the various parameters to be varied. The successful development of the baseline NGLV was assumed, leading to a baseline net cashflow from servicing the current satellite market. The fraction of the commercial market that the vehicle served could be varied, as could the size of the market.

From the baseline cash-flow, the NPV of the project could be calculated, using a required rate-of-return defined by the user. The present value of expansion options could then be calculated using the formulae given in Section 2.1 and a number of user-supplied variables such as the option cost (equating to expansion costs) and the volatility of the space tourism market. A number of consecutive options were considered, equating to the expansion required to serve the different market segments detailed in Section 2.2. Thus the first option was required to offer flights to customer group 1, the second to access customer group 2 and so forth. Again, the numbers and revenues generated by the differing customer groups could also be altered. In our baseline case, four consecutive expansion options were exercised 4,5,7 and 9 years after the initial investment year (2000), leading to the $50'000 per flight market. The exercise price of the expansion options was estimated through consideration of the fleet upgrades and new facilities required to access the market. The volatility of the space tourism market was assumed to be commensurate with similar high-risk future markets at $\sigma = 0.35$.

## 4. Results and Conclusions

The baseline scenario estimated the investment in NGLVs with a view to satisfying the satellite launch market to be marginally worthwhile. However, significant increments to the value of the investment were added by considering the options to expand into the space tourism market. In particular, the value of the last option to upgrade to the $ 50,000 per flight market provided a significant value increment. Further expansion into the 'mass' market may well prove even more attractive, though this must be balanced against the very high exercise price involved (i.e. the fleet/facilities required to service a market of 50,000 – 100,000 people).

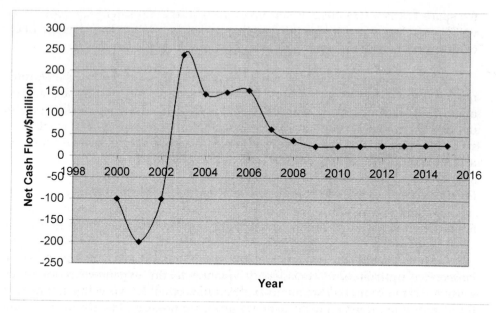

**Figure 2.** Cash flow for baseline case

| Investment stage | Option increment/$m | NPV/$m |
|---|---|---|
| 0 | - | 12.8 |
| 1 | 1.42 | 14.2 |
| 2 | 3.49 | 17.7 |
| 3 | 15.35 | 33.1 |

**Table 2.** NPV results for baseline investment
Investment stage 0 corresponds to initial investment with no real options.
Stages 1,2 and 3 factor in the real options of expanding into the first three
market stages of space tourism detailed in Section 2.2

The authors suggest that there is a window of opportunity, driven by the booming satellite communications industry, which should enable the development of reusable and semi-reusable launch vehicles with high levels of reliability. If the development of such vehicles takes too long then the current launch service providers, by providing the launch capacity through conventional means, may reduce the immediate attractiveness of developing a vehicle to cater for the LEO/MEO constellation niche. This will make the case for space tourism less clear as the investment in the launcher technology may well have to be aimed directly at this market, instead of first proving itself in the commercial satellite arena.

Currently the opportunity exists to profit from the launch of such constellations and, at the same time, provide valuable experience in the operational issues arising from a change to such a system. If the system developed exhibits goal congruence with the requirements of a vehicle to service the fledgling space tourism industry then the learning curve gradient, and thus the cost of exercising the option to begin tourism operations, will be significantly reduced. The return on investment should then be maximised for the investor.

## 5. Further Work

Preliminary results indicate that investment in NGLVs has the potential to be extremely profitable: this is due to the potentially huge market for space tourism being accounted for, rivalling and quite possibly exceeding the market for commercial satellite launches. What was once seen as a novelty is now seen by many to be the primary driver in the future industrialisation of space [Reference 12]. If this is indeed the case then the future of, and therefore investment in, NGLV projects will need thorough re-examination. In particular, those current projects that do not lend themselves well to expanding into a passenger –carrying role are in danger of locking themselves out of this revenue source.

The space tourism option does, however, have to be carefully managed. Development times in particular influence whether the investment is profitable or not. The suggested route to exploitation of this market is through the development of a reusable launch vehicle capable of meeting the needs of the present and near-term launch market (principally composed of telecommunications customers). Once this vehicle is established in the marketplace, incremental design improvement, further investment and benefits of the experience curve will allow the gradual incorporation of space tourism into the product mix. Another important point from this analysis is the consideration of the volatility of the space market: the near-term existence of such a market is subject to many political, economic and technological factors, so the volatility is high. However, this is an important reason why the option to enter such a market is worth so much; options allow full exploitation of upside potential whilst protecting against downside movement.

Those organisations that are first able to establish themselves as reusable launch vehicle providers will hold a very commanding position from which they may gradually exploit the new space tourism market.

The potential of using Real Options to provide a more long-term and accurate appraisal of the possible benefits of NGLV investment has been demonstrated theoretically. This analysis is by no means comprehensive: it should be noted that the baseline results presented here are preliminary and represent only one treatment relying on a number of assumptions and best-estimate values. Further work is being conducted with the aim of improving the accuracy of the analysis, along with studying the sensitivity of NGLV investment to the many parameters involved. A fully detailed and considered treatment of the subject would not only allow the feasibility of such investment to be proven, but also aid private NGLV developers to harness support from investors.

### References
1.     MCI Worldcom Corporate Web Site, <http://www.wcom.com/about_the_company/corporate_overview/internationa l_factsheet/>. *January 20, 2000*
2.     Qwest Corporate Web Site, <http://www.qwest.com/about/ir/faq>, *January 20, 2000*
3.     WTEC, 1998, *'Global Satellite Communications Technology and Systems'*, International Technology Research Institute, Loyola College
4.     Seynat, C., Smith, R., Hobbs, S., *Commercial Viability of Satellite Constellations as Future Earth Observation Systems*, RSS Conference UK, 1999
5.     Collins, P., Stockmans, R., Maita, M.: *Demand for Space Tourism in America and Japan, and its Implications for Future Space Activities* <http://www.spacefuture.com/archive...tourism_in_america_and_japan.shtml>. June 18, 1999
6.     Barrett, O.: *An evaluation of the potential demand for Space Tourism within the United Kingdom* <http://www.spacefuture.com/archive...sm_within_the_united_kingdom.shtml >. *June 16, 1999*
7.     Collins, P.Q., Ashford, D.M.: Potential Economic Implications of the Development of Space Tourism. *Ada Astronautica*, Vol 17, No.4, pp. 421A31, 1988
8.     Cheng F. Lee: Financial Analysis and Planning Theory and Application. Addison-Wesley Publishing company, 1985
9.     Brealey, R., Meyers, S.: *Principles of Corporate Finance*, 6th Ed. Irwin McGraw-Hill, 1996
10.    John Naisbitt, Global Paradox, Avon books, 1994
11.    Federal Aviation Administration, 1999,*'1999 Commercial Space Transportation forcasts'*, <http://www.ast.ffa.gov>. *September 17, 1999*
12.    Stephen Fawkes: Space Tourism – An end of Century Review, *Spaceflight*, Vol.42, February 2000

# A Review of Laser — Based Propulsion

**A. Pritchard,** International Space University, Strasbourg Central Campus, Parc d'Innovation, Boulevard Gonthier d'Andernach, 67400 Illkirch-Graffenstaden, France

e-mail: Pritchard@mss.isunet.edu, Spaceman95@Yahoo.com

**Abstract**

Laser-based propulsion is one of the more interesting ideas of 'engine-less' propulsion, where the mass penalties of traditional (chemical rocket) propulsion systems are avoided by eliminating the need for engine or fuel carried on board the spacecraft. In particular, compared with solar sailing, laser-momentum propulsion offers the possibility to gain orders of magnitude in photon pressure, and has therefore been proposed as a candidate for reaching large interplanetary and even interstellar distances. The solutions to the technological challenges posed to make such systems effective are non-trivial.

This poster paper was written during the author's Master of Space Studies placement at The Planetary Society, Pasadena, California. This paper discusses briefly laser-based propulsion, outlines some of the obstacles to laser sailing, and offers a view of the state of the art. A copy of a more detailed paper, written for the IAF Congress 2000 in Brazil, can be obtained on request from the author.

## 1.    Classification of Laser Propulsion

The energy of a laser beam can propel a spacecraft in several basic ways:

- Laser Sailing – similar principle to solar sailing, but a laser light source is coherent, generally more intense, and not subject to $1/R^2$ reduction like sunlight. Proposed by Tsander in 1924 [Reference 1], laser sailing history closely follows that of solar sailing, following Forward's work [Reference 2]

- Laser Thermo-Dynamic (LTD) / Laser Thermal Ablative – laser energy is focused in a reaction chamber to heat a propellant to produce thrust. LTD involves a gaseous propellant [References 3,4], thermal ablative vaporization of a solid fuel [References 3,5]. Extensively researched by military laboratories, with potential to benefit from industry spin -off/-in

- Laser Lightcraft – a ground-based laser beam, focused by the spacecraft body, ionises the air beneath the spacecraft. The resulting shockwave propels the craft. Propulsion of a small target several tens of metres vertically was already demonstrated [Reference 6], and there are plans to launch a 1 kg. satellite to orbit within three years [Reference 7 p17]

- Microwave Lightcraft –beamed microwave power ionises the air around a saucer shaped craft, and a Magneto-Hydro-Dynamic (MHD) motor

*M. Rycroft (ed.), The Space Transportation Market: Evolution or Revolution?*, 311–317.
© *2000 Kluwer Academic Publishers. Printed in the Netherlands.*

provides thrust [Reference 7 p18]. This innovation allows a LEO launch using either an orbiting power platform or a ground-based source

- Hybrid Solar Sail Assist – use of a laser to provide boost and high $\Delta v$ assistance to a solar sail.

## 2.    Missions for Laser Sailing

There are three distinctly difference classes of mission which could incorporate laser sailing.

- Extra-Solar Missions - A class of mission associated with both laser sailing and close solar flyby solar sails [Reference 8]. Exploration of the area surrounding the solar system includes numerous mission opportunities to explore the Heliopause at 100 Astronomical Units (AU) [References 9,10], the Kuiper Belt [Reference 10], the Sun's gravitational lens at 550 AU [References 9,10,11], the Oort Cloud [References 8,10], and the interstellar medium [Reference 10]

- Solar System Missions - Laser sailing is not considered as a competitive propulsion method inside the solar system [Reference 9]. This overlooks utilisation of laser sailing systems for fast reusable transportation, as part of evolving capabilities for longer and faster missions. NASA has discussed an Earth-Mars rapid-transit system using a 1GW accelerator/decelerator pair [Reference 10]

- Interstellar Missions - Associated with laser sailing due to the work of Forward [Reference 2] and Landis [Reference 12], the task of reaching the nearest stars is formidable – but laser sailing is perhaps the best technology to achieve this. Such an endeavour requires considerable long-term planning and development, as stated in the recent NASA-led Interstellar Science, Technology, and Missions Roadmap Study [References 7,8,10]. This also ties into Laser and Microwave Lightcraft technology in NASA's ASTP programme [Reference 13].

## 3.    Limitations to Laser Sailing

There are at least six major technological challenges to be overcome before a laser-propelled space mission of any significance can be realized.

- Laser Source - Current technology is limited to MW in flight [Reference 10] and 10s MW on the ground [References 7,10]. Achieving GW in orbit requires the same degree of technological development of laser technology as occurred in the last 20 years [Reference 10]. Recent investigations suggest that only 100s MW of microwave power may be needed [Reference 14]

- Optics - Focusing a high-power laser beam over a long distance is a considerable challenge. Forward [Reference 2] considers using a large Fresnel lens as the easiest method to achieve this but, as Landis [Reference 12] points out, this is a huge technological challenge. Alternatives include the phased-array aperture approach favoured by NASA [Reference 10], or the novel idea of the author to use a holographically projected lens

- Pointing - Pointing the propulsion beam at the sail is a large technological challenge. Clever use of Lagrangian points and/or solar-elevated non-Keplerian orbits [Reference 6], and optimum choice of optics aperture strategy can address this. Pointing accuracy for an Interstellar Mission, once thought to be impossible [Reference 10 p7], is less demanding than for the NASA (SIM) [Reference 15] and Terrestrial Planet Finder (TPF) missions [Reference 16] slated for launch in 6 and 11 years respectively. Navigation and control can be achieved using reconfigurable sails or movable centres of mass [Reference 17], and a beamrider design to remain on-beam

- The Space Environment - The environment of space itself is tough on sailing craft in particular – radiation can damage plastic sail substrates, high-velocity micro-impacts can lead to tearing, requiring additional mass for rip-stops in the sail design [Reference 17]. Microwave background and finite particle density limits the maximum velocity [Reference 18]

- Sail Material – The chosen sail material must survive high temperatures and the space environment, have good structural properties, and retain a low areal density. Previously best performance was obtained from metallic thin-films, with handling and construction difficulties [Reference 12]. Perforated sails reduce mass [Reference 19]; recently Carbon-Carbon micro-truss appeared as ideal sail material [References 7 p22, 20]

- Relativistic Effects – At the high velocities achievable using laser sails (one tenth of the velocity of light [Reference 10]) the Doppler effect leads to considerable loss of propulsion efficiency, as well as problems with communications. However this can be used to good effect to provide frequency-diplexed two-way communications.

## 4.    Case Study – Hybrid Solar Sail Assist in LEO

A solar sail in Earth orbit is limited to operation above a certain critical altitude. Solar radiation pressure alone is insufficient to overcome the effects of atmospheric drag below this point, which is variously cited as between 500 and 1000 km., depending on solar cycle and weather. Therefore solar sails generally require assistance to move beyond delivery to Low Earth Orbit typified by Space Shuttle, Mir and ISS orbits. It also takes a large amount of time for a solar sail to spiral out of low Earth orbit against drag and gravity

An orbital laser propulsion system can be a viable reusable alternative to chemical propulsion to solve these twin problems. This mission profile provides an excellent opportunity to deploy a first-generation in-space laser propulsion system on small scale, initially as a test-bed but possibly leading to use as an important infrastructure element for economical near-earth-orbit transportation. The system could also be used to provide higher thrust boost to assist with Δv manoeuvres not easily achieved by a solar sail alone.

Solar intensity at Earth Orbit is around 1.5 kW/m². To provide a similar intensity from a laser system, with efficiency of around 20% for a simple electrical pumped carbon-monoxide laser, suggests a spacecraft bus power requirement of order 10kW. The compact laser technology itself exists already – the US Airborne Ballistic Laser (ABL) system [Reference 10]. Pointing and control issues are more complicated, but the type of pointing control used on the Hubble Space Telescope (HST) should be sufficient. Laser recoil, for kW power levels and multiple-tonne platform, should not pose a significant additional problem. The relatively short operating range zone of the mission would not require a large or complicated optical aperture. Pointing would require an interlock and security to prevent misuse, although kW power levels do not pose a significant threat except perhaps to satellite optics. At kW output power levels, diplexing optical elements and imaging equipment, allowing the use of the platform as an imaging telescope, are not particularly complicated.

This sort of technology demonstration mission could be achieved by flying elements of laser technology developed for the US ABL programme, on a platform similar to the HST. Like HST, such a platform could also be serviceable, with next-generation technological improvements. This suggests that the required power levels and spacecraft design for a first-generation low-power assistant laser are achievable with the technology of today.

## 5.    Case Study – Optical Apertures and Novel Solutions

In standard treatments of laser sailing, collimation of laser beams over long distances requires large optical apertures. Forward [Reference 2], and others, consider a large Fresnel lens as the most achievable design. This requires a lens diameter of 100s-1000s Km to support interstellar flight, whilst at the same time the lens itself has to be extremely thin. As Landis points out [Reference 12], the extreme technological requirements of building such a structure, and maintaining precise control of its size, shape and orientation, are probably the most significant obstacle to realization of such systems. 'Traditional' laser sailing research concentrates on increasing maximum operating temperature and decreasing areal density of the sail, to allow a higher laser power over a shorter duration to achieve the same acceleration, reducing the beam collimation length and therefore lens size.

A more recent approach adopted by the NASA Interstellar Study [Reference 10] uses a large phased array of laser elements as a synthetic optical aperture for the propulsion system. Such a solution allows for a more flexible approach, and is more realistic with today's technology. It also allows a modular buildup of capability, and eliminates the complication of a three-body pointing problem. On the other hand, the mass-produced array elements require very precise positioning and constellation coordination.

The author proposes a third solution for consideration — the use of a virtual/holographically projected lens. This lens would be the interference pattern between two (lower power) laser sources, providing lensing of an incident propulsion laser beam by the same non-linear effect as observed in the micro-phenomenon of all-optical switching. If this technique, used commonly in telecommunications switching devices, is scaleable, it could provide significant reduction in complication of the laser-propulsion system.

An architecture of two interfering and one propulsion source reduces the mass, construction and positioning requirements of a laser propulsion system dramatically. Repointing the interfering beams can reconfigure the 'lens'. Two additional lasers increase the overall power requirements, but allow reuse of propulsion laser technology, and thus reduction in development costs. Reduction in optical beam-forming complexity also has large implications in optical communications, optical SETI, and interplanetary internet.

## 6.    Conclusions

The subject of laser sailing has been around almost as long as solar sailing, but for many years it was considered that the power levels were too extreme, the infrastructure requirements too great, and the gains too small compared to other more conventional systems, for this to be a viable technology in the foreseeable future. However, technology has a wonderful way of transforming the impossible into the practical. Novel approaches such as phased-array or holographic apertures, lasers with power levels achievable today, illuminating sails made of the latest Carbon-Carbon technology and carrying the latest Gossamer Spacecraft equipment, and with pointing and optical techniques developed in the search for extra-solar planets, laser sailing may come of age in the very near future. Following on from and intertwined with the development and imminent deployment of solar sails, laser sailing promises much as an alternative to chemical propulsion beyond near Earth orbit.

### Acknowledgements
The author thanks Dr Louis Friedman of The Planetary Society, for his support, advice, help and guidance during the difficulties of International Trade in Arms Regulations (ITAR) and Information Technology (IT) crises. Also he thanks Dr. Henry Harris of NASA who, despite remaining within the bounds of ITAR, managed to give the author a reinvigorating and inspirational view of a possible future, and Dr. Jim Burke, an unending source of inspiration and sound advice, who engaged himself tirelessly to ensure that MSS students on placement in California had the best possible experiences.

### References
1.    National Aeronautics and Space Adminstration (Tsander, K.): *From a Scientific Heritage*, NASA Report number TTF-541, 1967
2.    Forward, R.: Roundtrip Interstellar Travel Using Laser-Pushed Lightsails, *Journal of Spacecraft and Rockets*, Vol. 21, March-April, pp. 187-195, 1984
3.    Weiss, R., Pirri, A., Kemp, N.: Laser Propulsion, *Aeronautics and Astronautics*, Vol. 17, No. 3, pp. 50-58, March 1979
4.    Birkan, M.: Laser Propulsion: Research Status and Needs, *Journal of Propulsion and Power*, Vol 9, No. 2, pp. 354-360, March-April 1992
5.    Kantrowitz, A.: Propulsion to Orbit by Ground-Based Lasers, *Aeronautics and Astronautics*, Vol. 10, No. 5, pp. 74-75, May 1972
6.    CNN: *CNN Interactive Cold War Roadtrip*
      <http://cnn.com/SPECIALS/cold.war/experience/the.bomb/route/04.white.sands/>. Walton, A. – author, Atlanta, May 15, 2000
7.    Cole, J.: Really Advanced Propulsion Technology, *NASA Presentation to the American Institute of Aeronautics and Astronautics Florida Chapter*, January 2000.
8.    SPACE.com,    *NASA's    Vision:    Probes    at    Stars    by    2100*
      <http://www.space.com/news/21c-exploration_991231.html>. Clark, G. - author, December 31, 1999 (posted)
9.    McInnes, C.: *Solar Sailing: Technology, Dynamics and Mission Applications*, 1st ed., Springer-Praxis, Chichester UK, 1999

10. National Aeronautics and Space Administration: *Interstellar Science, Technology, Mission, Roadmap,* NASA Unnumbered Draft Document, 1999, Revision 0

11. Maccone, C.: (paper on proposed missions) *Acta Astronautica vol 43, nos. 9-10,* pp455-462, January-February 1999

12. Landis, G.: *Small Laser-propelled Interstellar Probe,* IAF-95-102, Proc. 46th International Astronautical Congress, Oslo, Norway, October 2-6, 1995

13. NASA, *Advanced Space Transportation Program Website,* <http://www.highway2space.com/>. Harris, D., May 16, 2000

14. Landis, G.: *Microwave-pushed Sails for Interstellar Travel,* Presentation at NASA 10th Advanced Propulsion Workshop, Huntsville, Alabama, April 5-8, 1999: <http://www.niac.usra.edu/files/studies/final_report/pdf/4Landis.pdf>.

15. Danner, R., Unwin, S.: *From Milliarcseconds to Micro-arcseconds,* <http://sim.jpl.nasa.gov/library/book/SIMBook_Chap1.pdf>. part of JPL SIM mission reference book, JPL, Pasadena, USA, May 15, 2000

16. Beichman, C. et al: *Probing the Central Arcsecond: General Astrophysics with TPF,* p73, <http://tpf.jpl.nasa.gov/library/tpf_book/Chapter_8c.pdf>. part of JPL TPF mission reference book, JPL, Pasadena, USA, May 15, 2000

17. Wright, J.: *Space Sailing.* Gordon and Breach Science Publishers S.A., Pennsylvania, USA, 1992

18. McInnes, C., Brown, J.: Terminal Velocity of a Laser-Driven Lightsail, *Journal of Spacecraft and Rockets,* Vol 27, No. 1, p48-52, January-February 1990

19. Matloff, G.L.: *An Approximate Heterochromatic Perforated Light-Sail Theory,* IAF-95-IAA.4.1.01, Proc. 46[h] Int. Astronautical Congress, Oslo, Norway, October 2-6, 1995

20. SPACE.com, *Space.com Exclusive: Breakthrough in Solar Sail Technology* <http://www.space.com/businesstechnology/technology/carbonsail_000302.html>. Clark, G. - author, March 2, 2000 (posted)

# System Analysis and Trajectory Optimization of an Ejector-Ramjet Powered Reusable Launch Vehicle

**R. Shepperd,** International Space University, Strasbourg Central Campus, Parc d'Innovation, Boulevard Gonthier d'Andernach, 67400 Illkirch-Graffenstaden, France

e-mail: shepperd@mss.isunet.edu

**A. Lentsch, M. Maita, T. Mori,** National Aerospace Laboratory, 7-44-1 Jindaiji-Higashi, Chofu, Tokyo 182-8522, Japan

e-mail: lentsch@nal.go.jp, mori@nal.go.jp, maita@nal.go.jp

**Abstract**
The ejector ramjet, one kind of rocket-based combined cycle propulsion, has received increasing attention lately as a potential solution for reusable air-breathing space transportation systems. This paper discusses a system study of an ejector ramjet and presents a method for the optimization of the trajectory and engine parameters. Due to the potentially large number of input parameters and their sensitivities on the overall system performance, a hybrid genetic algorithm was chosen for the optimization.

## 1. The Ejector Ramjet

In conventional oxygen/hydrogen fueled rockets, the oxygen represents 80 to 85% of the propellant mass. Using atmospheric oxygen during the ascent thus offers an attractive possibility to lower the propellant mass fraction, which is crucial to realizing reusable, particularly single-stage, launch vehicles. An ejector ramjet, the simplest solution in the large family of air-breathing combined-cycle engines, uses a rocket jet to draw air into the engine where it is compressed, heated, and expanded [References 1, 2]. In this study, a thermodynamic engine model of an ejector ramjet is simulated over a 2-dimensional ascent trajectory to identify basic design aspects and sensitivities.

## 2. Optimization and System Analysis

Traditional calculus-based optimization tools, though efficient, lack the ability to identify global solutions or handle noisy or discontinuous functions. Moreover, parameters for launch vehicle trajectories often exhibit a high degree of sensitivity, particularly with the inclusion of engine parameters for optimization and system analysis. The ejector ramjet model considered here displays all of these drawbacks and eludes optimization by gradient methods. The model's advantage, however, is its quick execution time, allowing several thousand evaluations to be made in a couple minutes on modern workstations.

319

*M. Rycroft (ed.), The Space Transportation Market: Evolution or Revolution?*, 319–323.
© 2000 *Kluwer Academic Publishers. Printed in the Netherlands.*

Optimization via genetic algorithms was therefore chosen and hybridized with the robust downhill simplex method of Nelder and Mead [References 3, 4].

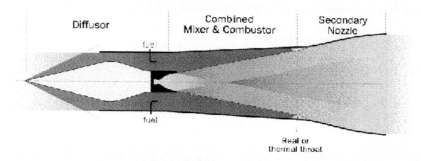

**Figure 1.** Ejector ramjet

## 3. Results

The genetic algorithm used in this study employs bit strings with 8 bits of resolution per argument, Gray encoding, windowing fitness, roulette wheel selection with elitism (the best 2 chromosomes are copied into the next generation), multi-point crossover, and bit-flip mutation. Good results were obtained with a population size of 100, a crossover probability of 80%, and a bit-flip mutation probability of 1%. Fig. 2 shows the performance history of a typical run. The initial rapid rise is due to the crossover operator, where the genes of the population are re-combined to create better children.  In the shallower performance rise of the latter generations, the population has begun to converge and mutation plays the dominant role. The downhill simplex method is applied here after the population has converged. If no local optimum exists between this point and the global optimum, the downhill simplex method can obtain and refine the optimum solution faster than mutation alone, reducing the total number of functional evaluations.

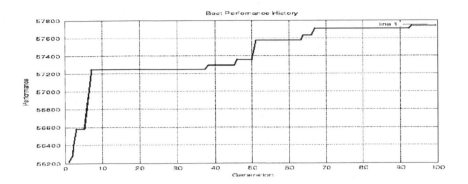

**Figure 2.** Performance of the Ejector Ramjet model in units of km²/sec (specific angular momentum of the final circular orbit) as a function of generation (100 evaluations per generation)

Successful global optimizations of the model have been performed and the algorithm is now aiding engine configuration studies with an emphasis on the ejector mode.[1] For the rocket engine, higher combustion chamber pressure and chamber temperature increase ejector mode performance. A chamber pressure of $p_c$ = 20 Mpa and chamber temperature of $T_c$ = 3500 K was therefore used. Higher values of secondary mass flow are also better. For the rocket engine's oxygen to fuel ratio, deviations from the optimum value tend to decrease the overall ejector performance, so common values were selected (Oxygen/Fuel = 5.3 - 6.0, depending on the chamber pressure). Secondary fuel injection into the unmixed airflow did not show a performance increase and was therefore dismissed. A slightly increasing mixer cross-sectional area (outflow/inflow area of 1.2) resulted in the best trade between the combustion pressure losses and the thermal choking limit. In the ejector mode, the engine was operated at the thermal choking limit at the mixer exit, which showed the best performance. With these settings, the ejector mode resulted in approximately a 15% increase of specific impulse at takeoff, a 35% increase (600 seconds) at Mach 2.0, and an impulse of 700 seconds around Mach 4.[2]

---

[1] Ejector mode runs between takeoff and about Mach 2, when Ramjet operations begin
[2] An upper limit for the case when the Ramjet is not used

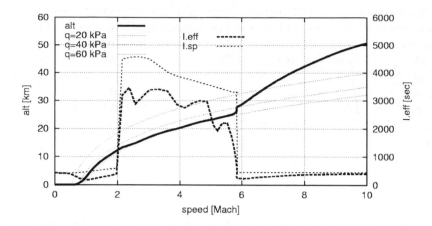

**Figure 3.** Ascent trajectory for a horizontal takeoff single stage to orbit vehicle with a combined cycle propulsion system: the altitude, specific impulse, and the effective specific impulse (net thrust divided by the fuel mass rate flow) are plotted against speed

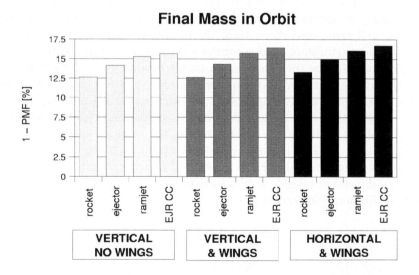

**Figure 4.** Vehicle concept comparison

To assess the overall benefit of the propulsion system, three different vehicle concepts were analyzed: a conventional vertical takeoff rocket with no lift, a vertical takeoff vehicle with wings for extended atmospheric flight, and a horizontal takeoff vehicle with low Lift/Drag characteristics (a winged rocket shape). These three vehicles were simulated with 4 different propulsion concepts: rocket only, ejector rocket, rocket-ramjet, and ejector-rocket-ramjet combined cycle. The resulting final mass (1 minus the payload mass fraction) is shown in Fig. 4. The genetic algorithm was used to optimize 14 parameters controlling the trajectory. In all cases, the combined cycle shows a clear performance gain at a potentially low cost of additional mass.

### References

1.  Escher, W. J. D., Hyde, E., Andserson, D., *A User's Primer for Comparative Assessments of all Rocket based Combined Cycle Propulsion Systems*, AIAA 95-2474, 1995
2.  Lentsch, A.: *Potential Benefits and Limitations of Ejector Ramjets for Reusable Launch Vehicles*, ISTS 2000-g-13, 22nd International Symposium on Space Technology and Science, Morioka, Japan, 2000
3.  Davis, L. (editor): *Handbook of Genetic Algorithms*. Van Nostrand Reinhold, 1991
4.  Shepperd, R.: System Optimization of an Ejector Ramjet Rocket-Based Combined-Cycle Model, MSS Thesis. International Space University, 2000

# A Launch Operations and Capacity Model of the U.S. Eastern Range

D. Steare, Lean Aerospace Initiative, Massachusetts Institute of Technology, 77 Massachusetts Avenue, Room 41-205, Cambridge, MA 02139, USA

e-mail: steared@mit.edu

### Abstract

This paper describes the methodology used to create a computer model for examining current and future launch operations of the U.S. Eastern Range. It also explains the results produced by running multiple simulations with the model. Space launch, like all other modes of transportation, is valuable to a broad spectrum of users and serves the interests of civil, commercial, and national security communities. As spaceport traffic increases, estimating launch capacity and determining a proper pace of operations become critical issues for both launch program planners and range operators. An expanding launch market with increasing priorities on the cost of space access has created a greater demand for lean operations of launch-support facilities. The Eastern Range of the United States is the busiest launch facility in the world and supports a host of launch vehicles with varied characteristics. The modeling approach utilized the field of System Dynamics to incorporate physical constraints, personnel limitations, program requirements, projected conditions, and various other factors that impact launch operations. Although the modeling effort concentrated on a specific launch range, its general cause and effect relationships could be reapplied to any launch facility. Such a model is a useful tool for better understanding and recognizing the interactions among multiple launch programs utilizing resources of a common facility. It can also be used to simulate scenarios of future operating environments and policies to aid in the upgrading or planning of new launch facilities.

## 1. Background

The Eastern Range of the United States is comprised of the Kennedy Space Center and Cape Canaveral Air Force Station as well as several down range support stations. The entire launch network constitutes the busiest launch facility in the world. This national range supports the operations of multiple launch vehicle programs, each with unique support characteristics. In early 1998, the 45th Space Wing of the United States Air Force (USAF) undertook a range capacity modeling effort. This resulted in the creation of a first-generation computer model capable of producing annual values for Eastern Range launch capacity. Subsequently, the model was adopted by the USAF and has contributed to a number of recent launch-related studies including: the *Range IPT Report* [Reference 1] and the Congressional *National Launch Capabilities Study* [Reference 2].

The USAF range capacity model was developed using common spreadsheet software and has undergone periodic updates since its creation; however, its primary structure and capabilities have not fundamentally

*M. Rycroft (ed.), The Space Transportation Market: Evolution or Revolution?*, 325–327.
© 2000 *Kluwer Academic Publishers. Printed in the Netherlands.*

improved during the past two years. In the fall of 1999 an independent modeling project began at the Massachusetts Institute of Technology (MIT), as part of the ongoing Lean Aerospace Initiative program, in order to explore the possibility of applying a system dynamics[1] approach to the problem.

## 2.    Modeling Approach

The system dynamics model utilizes much of the work and many key variable relationships already applied to the existing USAF model. The underlying assumption of both models is that the possible number of annual launches from a single range is a function of the range support requirements for each proposed launch vehicle campaign and the ability of the range to provide this support for each campaign. The fundamental unit of measure chosen to model launch capacity was time.

Launch capacity is derived from the time requirements of the following cumulated factors: launch vehicle operations support durations, effects of scheduling changes, impact of range systems maintenance, planned range downtime, and range personnel workload limitations. The system dynamics model goes beyond the capabilities of the USAF deterministic approach. Rather than simply calculating a single value for the maximum possible number of annual launches, this advanced methodology uses probabilistic methods to determine an entire set of likely scenarios. All of these simulations are then combined to produce a distribution of expected launch totals.

The model is defaulted to simulate the course of activities for a given year on an hour-by-hour basis. For each hour of a simulated year the model accomplishes the following sequential tasks: 1) determines the current time, 2) checks range status for operations restrictions, 3) calculates the need for range crew rest, 4) accounts for support rescheduling effects, 5) performs maintenance on range systems as necessary, and 6) prioritizes and allocates range support for launch operations.

The layout of the model has been organized into multiple sections for ease of understanding and future enhancement including: time functions, planned restricted days, crew rest, range systems maintenance, model output and statistics, launch vehicle operations, scheduling impacts, prioritizing support requests, and range support allocation. The current version of the model, a relatively large application of system dynamics, can readily accommodate up to

---

[1] System dynamics is a modeling methodology founded by Jay Forrester at MIT in 1956 that combines theory and computer simulation while accounting for internal feedback-loop relationships

14 simultaneous launch campaigns. The model is considered to be a relatively large system dynamics model. It accounts for approximately 70,000 unique variables. Running a single hour-by-hour simulation of a given year requires the calculation of roughly 0.6 billion explicit values.

As an example, the system dynamics model was used to run a series of 50 simulations of the Eastern Range manifest from the fiscal year 2001 National Launch Forecast (NLF) (*September 1999 edition*). The expected number of launches resulting from these simulations is displayed in the following figure.

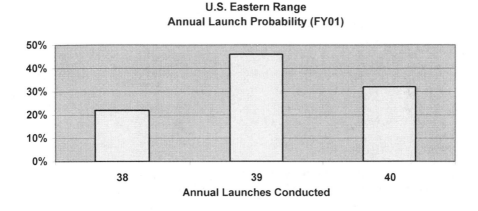

**U.S. Eastern Range**
**Annual Launch Probability (FY01)**

**Figure 1.** Example of model output (expected launch total for fiscal year 2001)

## 3.    Conclusion

The system dynamics approach has been successfully applied to enhance the capabilities of an earlier approach to modeling the capacity of a rocket launching facility. Using this advanced methodology, increased operations fidelity has been achieved. As a result, such a modeling tool offers the opportunity for significantly improved understanding of the complex interactions among multiple launch programs utilizing the resources of a single facility. This approach offers much aid in simulating scenarios of potential operating environments, and also in evaluating possible concepts to upgrade and plan future launch facilities and spaceport operations at these facilities.

References
1.    U.S. Air Force: *Range Integrated Product Team Report*, March 10, 2000
2.    U.S. Department of Defense: *National Launch Capabilities Study*, July 8, 1999

# An Academic Perspective to Revolutionize the Manned Space Transportation Market

**W. White,** Ryerson Polytechnic University, Department of Mechanical Engineering, Toronto, Ontario, Canada

**W. Brimley,** Ryerson Polytechnic University, School of Aerospace Engineering, Interactive Learning Connection - University Space Network, Toronto, Ontario, Canada

e-mail: wwhite@acs.ryerson.ca, bbrimley@acs.ryerson.ca

**V.J. Lappas,** University of Surrey, Surrey Space Center, Guildford, Surrey, UK

e-mail: v.lappas@ee.surrey.ac.uk

### Abstract

The present paper will give an academic view of how a team of 18 undergraduate students approached the topic of the Manned Space Transportation Market. It will describe the design of a (SSTO), Space Transportation System undertaken as part of their senior undergraduate thesis course in Spacecraft Systems Design, at Ryerson Polytechnic University, Toronto, Canada. Their design was in part a response to the challenge of the 1st X-Prize University Competition. Not only did the student team perform the preliminary design of a manned sub-orbital vehicle, but by also using the concept of modularity they were able to extend its capabilities to orbital flight.

The student team went on to present their design at M.I.T. in the spring of 1998. The Ryerson Team exercise was also duplicated at Queen's University, Royal Military College, and York University, since the Spacecraft Systems Design Course is run concurrently on the Internet between these Canadian universities. Each of the other university teams developed their own design to meet the requirements laid down by the X-Prize design competition. The viability of the designs using existing technology was proven.

This student exercise provides insight and answers, from the academic point of view, to these questions raised in the space transportation sector:

- Re-usable vs. expendable space vehicles
- Government vs. private sector design and space operations
- Use of existing launch and landing facilities
- Commercialization of Manned Space Flight.

It is the intention of this paper to add a constructive, academic perspective which supports the need to revolutionize the Manned Space Transportation Market.

## 1. The Ryerson X-Prize Vehicle System (XVS) 'Space Plane'

The XVS vehicle is a Single Stage To Orbit (SSTO) spaceplane with a Vertical Take Off-Horizontal Landing (VTOHL) capability. It was designed as part of the Spacecraft Systems Design course, run at Ryerson Polytechnic University, in conjunction with the University Space Network. Its mission statement is similar to that outlined by the X-Prize Organization: Design a

*M. Rycroft (ed.), The Space Transportation Market: Evolution or Revolution?, 329–331.*

reusable manned space vehicle, able to reach an altitude of 100 km (suborbit), safely return its 3 passengers with the condition of not replacing more than 10% of the flight vehicle's first-flight non-propellant mass between two flights and by repeating the same mission twice in 14 days. The XVS vehicle was not only designed to meet these requirements but also to meet an orbital requirement and 100% reusability.

However, designing a spacecraft to meet both suborbital and orbital requirements is almost like designing two different spacecraft with different physical properties (mass, fuel, power, volume, thrust). Presented with this dilemma, the Ryerson X-Prize team utilized the idea of developing a modular type space plane which consists of three parts: the common environmental capsule (containing the passengers, pilot and environmental systems) and the suborbital and orbital modules (which would each contain the necessary fuel and number of engines depending on the mission). Table 1. indicates the mass and size properties of both vehicles.

| Physical Properties | Suborbital | Orbital |
|---|---|---|
| Total Mass | 58,648 kg | 145,260 kg |
| Engines | 1xRS-2200 | 3xRS-2200 |
| Thrust (S-L) | 206,500 lbf | 619,500 lbf |
| Length | 17,5 m | 22 m |
| Width | 13 m | 16 m |
| Height | 7.3 m | 7.9 m |
| Wing Surface | 15 m² | 18 m² |
| Rudder (Height) | 1.9 m | 3 m |

**Table 1.** Physical properties of XVS suborbital and orbital cases

The work completed is impossible to be presented in a two-page paper. Eighteen students put their best effort into designing a Manned Space Transport System with very challenging mission requirements and resources. The outcome was the preliminary design of an innovative SSTO spaceplane. From an academic perspective, it supports the X-Prize proposal to the space community and academia of how the Manned Space Transportation Market should change in order to provide a better, affordable and more frequent opportunity to reach space. A more extensive and detailed report on the XVS design can be found in the references below (including the X-Prize University competition presentation), as well as the topics of 18 individual student theses related to the XVS design.

**Figure 1.** XVS in ascent stage and XVS suborbital isometric view

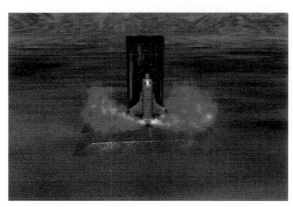

**Figure 2.** XVS at the launch facility and XVS with landing gear and flaps
extended

**References**
1.  Lappas, V.J.: *XVS Mission Operations Analysis,* AER080 B.Sc. Thesis, Ryerson Polytechnic University, Toronto, 1998
2.  Lappas, V.J., Wilson, A., Lardizabal, E., Birket, J.,.: *Ryerson XVS Presentation,* X-Prize University Competition, MIT Mass., 199
3.  University Space Network (USN): *Ryerson X-Prize Proposal,* <http://www.ilc-usn.kcc.ca/default.html>[1]. April 20, 2000Toronto, May 3, 1996
4.  X-Prize Foundation, < http://www.xprize.org>. April 20, 2000

---

[1] The website address for the ILC-USN is to be moved during summer 2000 to <www.ilc-usn.net>